Studies in Computational Intelligence

Volume 1000

The series "Studies in Computational Intelligence" (SCI) publishes new developments and advances in the various areas of computational intelligence—quickly and with a high quality. The intent is to cover the theory, applications, and design methods of computational intelligence, as embedded in the fields of engineering, computer science, physics and life sciences, as well as the methodologies behind them. The series contains monographs, lecture notes and edited volumes in computational intelligence spanning the areas of neural networks, connectionist systems, genetic algorithms, evolutionary computation, artificial intelligence, cellular automata, self-organizing systems, soft computing, fuzzy systems, and hybrid intelligent systems. Of particular value to both the contributors and the readership are the short publication timeframe and the world-wide distribution, which enable both wide and rapid dissemination of research output.

Indexed by SCOPUS, DBLP, WTI Frankfurt eG, zbMATH, SCImago.

All books published in the series are submitted for consideration in Web of Science.

Aboul Ella Hassanien · Mona Soliman
Editors

Artificial Intelligence: A Real Opportunity in the Food Industry

Springer

Editors
Aboul Ella Hassanien
Faculty of Computers and Artificial Intelligence
Cairo University
Giza, Egypt

Mona Soliman
Faculty of Computer and Artificial Intelligence
Cairo University
Giza, Egypt

ISSN 1860-949X ISSN 1860-9503 (electronic)
Studies in Computational Intelligence
ISBN 978-3-031-13701-3 ISBN 978-3-031-13702-0 (eBook)
https://doi.org/10.1007/978-3-031-13702-0

This Springer imprint is published by the registered company Springer Nature Switzerland AG
The registered company address is: Gewerbestrasse 11, 6330 Cham, Switzerland

Preface

The concept of food quality is based on a complex and multidimensional concept influenced by a wide range of situational and contextual factors. The emerging technology, led by artificial intelligence, has affected all areas of life, such as food quality, where data will play an essential role in the development of the food industry. IoT sensors directing irrigation to the correct areas reduce excessive water use and subsequent environmental impacts such as agricultural chemical intrusion and soil erosion. Moreover, it can reduce the water consumption and labor needed for farming. This improves efficiency and provides high-quality food resources that meet the Sustainable Development Goals. AI applications will impact food trading establishments and help them build income and elevate the customer experience. Moreover, shoppers can access many food-focused AI stages.

This book focuses on applying artificial intelligence in the sustainable food industry. It contains nine chapters. The first chapter presents a method that assimilates RGB information with time-of-flight depth data using a Kinect2 sensor to create accurate 3D models of the trunk and primary branches of dormant apple trees. The second chapter aims to detect various types of leaf diseases to differentiate them from healthy crops without needing continuous monitoring and minimizing yield losses. To increase the efficiency of post-harvest industrial processes, there is a need to introduce technologies for classifying different species of pistachio. So, the third chapter proposed an automated model to separate pistachio species. While the fourth chapter presents a unique model for classifying rice leaf diseases by multi-class support vector machines and K-means clustering with particle swarm optimization. The fifth chapter deliberates some applications of AI, primarily focused on convolutional neural networks (CNNs) currently being used in the food industry, challenges confronted in the present-day technology, and most notably, some of the future applications of CNN in the food industry. The sixth chapter discusses some applications of AI-based food quality. The seventh chapter attempts to use different deep learning architecture approaches to spot this illness in coffee leaves and develop an expert system to help coffee farmers diagnose problems early on. The eighth chapter, setting up a system for monitoring seawater quality to help control Fiji water pollution using IoT. An Internet of Things (IoT)-based framework for food monitoring is proposed in the ninth chapter to protect food from spoilage caused by changes in environmental conditions during storage. In the current scenario, the prediction was made using recorded sensed data, and a detailed analysis was performed to identify the factors causing the food to spoil. This chapter proposes an automated controlling mechanism for environmental parameters using adaptive random forest prediction and IoT.

Giza, Egypt — Aboul Ella Hassaien
Giza, Egypt — Mona Soliman

Contents

Application of Artificial Intelligence in the Food Industry: AI-Based Automatic Pruning of Dormant Apple Trees

Noha Elfiky

Abstract Automatic pruning is one of the most essential as well as expensive and labor-intensive procedure in specialty crop production. During winter, expert pruners are employed to remove certain primary branches of dormant trees based on a set of predefined rules. The end goal of the automatic pruning applications is to reduce dependence on this huge labor and associated humongous costs by automating the pruning decisions, which in turn would facilitate easy interface with intelligent robotic pruners. In this chapter, we have developed a robust *3D* reconstruction scheme that utilizes color information in addition to the time-of-flight depth data from *Kinect2* sensor for accurately modeling the trunk and primary branches of a dormant apple tree. The motivation behind incorporating color in our framework is to acquire accurate *3D* tree reconstruction even in absence of quality depth data. Quantitative and qualitative comparisons of our proposed approach with a depth based reconstruction technique depicts the usefulness of using color information in our automatic pruning system. On average, the proposed scheme provides a performance of 93.94% accuracy for correctly identifying the branches, and 71.13 and 89.26% for correctly estimating the diameters of the primary branches within error margins of 3 mm and 5 mm respectively. Moreover, the time complexity of our proposed algorithm shows an improvement in order of magnitude over the complexity of the baseline approach.

Keywords Intelligent computing · Food industry · 3D modeling of dormant apple trees · RGBD data · Artificial intelligence · Big data analysis

1 Introduction

Dormant pruning, an essential procedure in specialty crop production, is necessary for healthy growth of trees and better yield of fruits. It basically refers to the process of removing certain primary branches (i.e., branches that are connected directly to the

---

N. Elfiky (✉)
Saint Mary's College of California, Moraga, CA, USA
e-mail: nme5@stmarys-ca.edu

A. E. Hassanien and M. Soliman (eds.), *Artificial Intelligence: A Real Opportunity in the Food Industry*, Studies in Computational Intelligence 1000,
https://doi.org/10.1007/978-3-031-13702-0_1

trunk) of leafless trees which do not produce fruits anymore. During winter, a large manpower is recruited for this pruning purpose. It is regarded as one of the costly and labor-intensive processes in fruit production. To reduce the dependence on huge labor and associated costs, researchers are now working extensively towards mechanizing the entire pruning process. Precise geometric information about the trunk and primary branches of the trees is required for correct identification of the potential pruning points, the final goal being supplying all this information to an intelligent robotic agent for automatic pruning. Thus, 3D reconstruction and modeling of the trees with accurately measured trunk and branch diameters has become one of the crucial stages of this pruning process.

The aim of this study is to propose an RGBD based 3D reconstruction technique for creating robust and accurate models of dormant apple trees. Most previous works [2, 7, 10, 17] have used only depth images for tree modeling purpose. The contribution of this chapter is the significant improvement in runtime complexity and reconstruction accuracy over other baseline approaches by integration of color information in our framework. Literature on feature extraction and matching in color images is vast. So, we expect to extract better and more robust features from RGB images, and thus design a more accurate alignment framework in our system.

Moreover, the Kinect2 sensor does not have a very high depth resolution (425×512), as it has been designed to be used for indoor gaming interface. Thus, when used outdoors, quality of the depth data from Kinect2 might not be perfect, and may have missing pixels. The color features can be used in those cases to fill in the missing data with an expectation of achieving better results. The proposed algorithm has a multi-layered structure, the first layer being the preprocessing framework where we create semicircle fitted models for each depth image of a tree, as in [2, 7]. The second layer involves creating RGBD frames from corresponding pairs of color and depth images as shown in Fig. 1, and the final layer is our three stage registration framework where initial transformation is obtained from a consecutive pair of RGBD image frames as in Fig. 2a, the second stage of transformation is achieved by aligning the corresponding pair of entire 3D point clouds as in Fig. 2b, and finally, a fine tuning of alignment is performed on the pairs of semicircle fitted models whose 3D point clouds are found to be similar as displayed in Fig. 2b. Once corresponding 3D point clouds are aligned, a globally consistent merging technique is employed to accumulate all these data for creating the final 3D model of an apple tree.

The remainder of this chapter is organized as follows. An overview of the current state-of-the-art is presented in Sect. 2. Section 3 explains our preprocessing pipeline and final 3D reconstruction technique. An in-depth description of the experimental set-up containing the dataset details and the evaluation metrics is given in Sect. 4. Experimental results are discussed in Sect. 5. Finally, in Sect. 6 the content of the chapter is summarized and potential future works are highlighted.

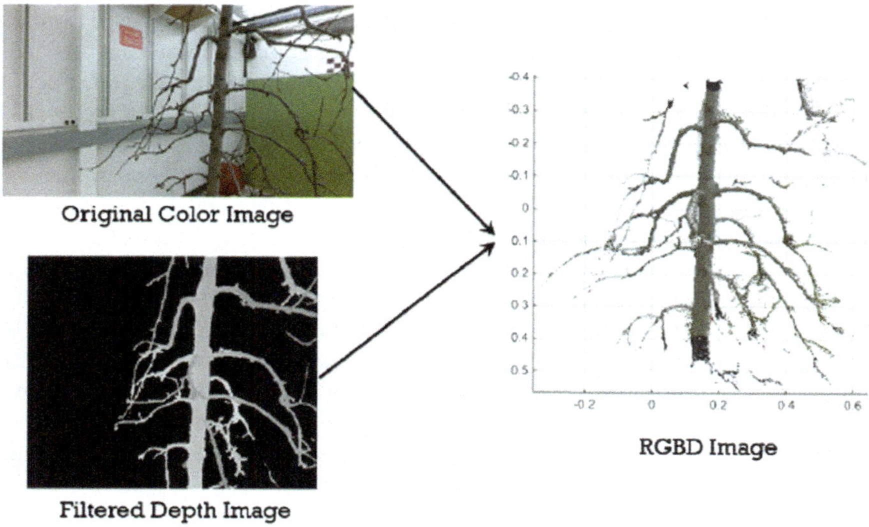

Fig. 1 RGBD frame created from a corresponding pair of depth and color images

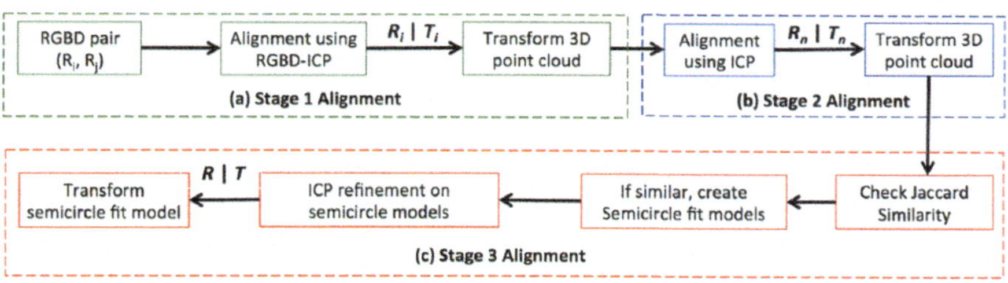

Fig. 2 Cutting the primary branch points cloud out of the original tree points cloud

2 Related Work

The robotics and the computer vision communities have a vast and matured literature on geometric reconstruction and 3D scene modeling. This section provides a brief overview of the state-of-the-art of 3D reconstruction and modeling techniques in the field of automatic pruning of trees. We mainly focus on the Kinect Fusion based approaches which make effective use of RGBD data and 3D point clouds for 3D reconstruction and modeling.

The past few decades have seen an increased interest in 3D technologies such as stereo vision, laser scanners and structured light for 3D sensing and object reconstruction. However, only limited studies aimed at recreating 3D appearance of trees. Moreover, the literature mostly focused on reconstructing trees with foliage [18, 23].

In [26], a shape-from-silhouette based tree reconstruction scheme is used to model dormant trees in indoor setup. However, performance of the technique in actual high density modern orchards is not confirmed. An RGBD sensor based 3D reconstruction algorithm for robotic pruning of dormant trees is proposed by Wang et al., in [29]. Though their results look promising in a laboratory setup, the accuracy deteriorates in outdoor orchard environment in presence of direct sunlight. Other popular methods [1, 17] employ generative branching rules without taking into account the fact that the trees are predominantly cylindrical in structure, and thus using this information may lead to simplification of the tree structure. These methods may give inaccurate reconstruction for complex tree structures in the absence of large number of hypotheses. Another method [19] focuses on reconstructing complex trees from a sparse set of images by integrating 2D/3D tree topology as shape priors into the modeling process. Some of the previous methods have also aimed at modeling dormant trees using stacked circles [10] and semicircle fitted 3D reconstruction schemes [2, 7], using only depth imaging information, and have achieved pretty promising results. However, their results suffer when there are missing depth data, or when the data is collected in direct sunlight.

Most reconstruction systems need accurate spatial alignment over consecutive data frames, and globally consistent alignment of all data frames. Thus, the registration of point clouds to obtain an accurate reconstruction of the trees is an important stage in our algorithm. Several methods exist in literature to find the full 6DOF scan alignment, such as joint optimization combining visual features and shape matching [13, 14, 33, 34], overlap heuristic approach [30], semidefinite programming [8], probabilistic approach [9] that takes into account the uncertainty associated with displacement of the sensor and the measurement noises, and many more [5, 27]. The most important and oldest class of alignment algorithms has been based on the Iterative Closest Point Algorithm (ICP) [4], a nonlinear optimization problem where correspondences between images are approximated using the closest pairs of points found between scans at the previous iterations. The solution to the frame alignment problem strongly depends on the data being used. In this chapter, we have used an advanced version of ICP (known as RGBD-ICP) [14] designed to register consecutive RGBD data frames.

All the images used for our experiments are acquired with a Kinect2 sensor using its Kinect Fusion software [15]. We have chosen Kinect2 for our research because of its low cost and easy portability. Although other laser scanners like LIDAR provide images of better quality, they bear the side effects of being extremely expensive and heavy. From our experiments, we could see that the Kinect2 sensor is capable of providing us with depth images of comparable quality as of other expensive sensors. Moreover, we could collect the data at a really high frame rate of 30 fps. Owing to these benefits, several other traditional approaches [20–22, 25, 32] have also used Kinect for data capture.

3 The Proposed Methodology

In this section, our entire 3D reconstruction framework is explained in details. The preprocessing operations in the proposed pipeline, which includes sensor calibration, noise filtering, and closing holes in the raw data is discussed in Sect. 3.1. Our adopted technique [2, 7] for creating semicircle fitted models from depth images is explained in Sect. 3.2. Then, in Sect. 3.3, we explain how to create a 6-dimensional RGBD frame from a pair of depth and color image. Finally, Sect. 3.4 describes the proposed 3D reconstruction scheme.

3.1 Preprocessing Framework

(1) ***Sensor Calibration***: As the final goal of the proposed pipeline is to achieve better reconstruction results by incorporating color information in our pipeline, we need to have the depth and color information in the same frame of reference, aligned on top of each other. Thus, calibration of the sensor becomes a crucial stage in the pipeline.

The first step in our preprocessing framework, thus, involves calibrating the Microsoft Kinect2 in order to remove the lens distortion from depth as well as color images. Our application requires a perfect knowledge of the intrinsic calibration parameters (focal length, principal point, radial, and tangential distortion coefficients) of the sensors for an accurate 3D reconstruction [16]. For this purpose, the single sensor calibration is done based on classical checkerboard corner detection algorithm using GML C++ toolbox [28]. The depth sensor parameters are computed using the infrared images; as they come from the same sensor of the Kinect device.

The intrinsic calibration matrix of a sensor is defined as follows:

$$\boldsymbol{K} = \begin{bmatrix} f_x & s & c_x \\ 0 & f_y & c_y \\ 0 & 0 & 1 \end{bmatrix} \tag{1}$$

where (f_x, f_y) are the focal lengths (in pixels) among both image axes, $\left[C_x, C_y\right]^T$ represents the principal point, and s is the skew factor.

The intrinsic parameter vector k is obtained by calibrating the sensor. Subsequently, the undistorted depth map can be obtained as follows:

$$X_t = \begin{bmatrix} 2k_3xy & k_4\left(r^2 + 2x^2\right) \\ k_3\left(r^2 + 2y^2\right) & 2k_4xy \end{bmatrix} \tag{2}$$

$$X_u = \left(1 + k_1r^2 + k_2r^4 + k_5r^6\right)\boldsymbol{X} + \boldsymbol{X}_t \tag{3}$$

where $X = (x, y)$ is the input depth map, X_t is the tangentially undistorted map, X_u is the radially and tangentially undistorted map, $r^2 = x^2 + y^2$, and $k = (k_1, k_2, k_3, k_4, k_5)$ is the parameter vector, where k_1, k_2, and k_5 are the radial distortion parameters; while k_3 and k_4 are the tangential distortion parameters.

To incorporate color information in our framework, we need to have knowledge about the rigid transformation between the sensors of Kinect2. Thus, stereo calibration is performed to find the extrinsic parameters of the device. The extrinsic parameters give the orientation of the cameras w.r.t. a chosen frame of reference. It includes a 3×3 rotation matrix (R) and a 3×1 displacement vector (t).

(2) ***Filtering Noise from Raw Depth Images***: A substantial portion of any raw RGBD frame is occupied by the background. Once lens distortion is corrected, the next step in our algorithm is to remove most of the background from the depth images by taking only the points within a 2D bounding box where we expect to find the tree object, as in [7].
(3) ***Filling up the Gaps in the Depth Images***: Another challenge in accurate 3D reconstruction of the dormant apple trees is the missing data i.e., the presence of gaps and holes in the depth images. To overcome this problem, we apply grayscale morphological operations to fill up the internal gaps present in the images, as in [7].

3.2 Creating Semicircle Fit Model

Once we have a filtered clean depth image, we first extract the tree skeleton from the depth image by employing a Voronoi diagram based technique, as proposed by Yohai et al., in [24]. As output of skeletonization, we receive the 2D skeleton nodes and their interconnections. We then aim to find the pair of edge points surrounding each skeleton node. We refer to the region surrounding a particular skeleton node along with the pair of edge pixels as the cross-section of that skeleton node. Our next goal is to find the center and the radius of each cross-section of the tree. In our proposed technique, we have used a diameter correction model, as suggested by Shayan et al., in [2], to find the radius r of each cross-section. The corresponding center can be then estimated as follows:

$$C = \mu + \boldsymbol{r}\upsilon \tag{4}$$

$$S_{3D}(n) = \mu \mathrm{K}^{-1} \boldsymbol{S}_n \tag{5}$$

$$\upsilon = \frac{S_{3D}(n)}{S_{3D}(n)} \tag{6}$$

where S_n is the nth skeleton node with a depth value μ, K is the camera intrinsic matrix, $S_{3D}(n)$ is the 3D point corresponding to S_n, and υ is the Kinect Observation

Vector (KOV) at S_n. Finally, the optimal semicircle points at radius r can be fitted around the center C by employing the Rodrigue's formula, as in [11].

3.3 Creating RGBD Frames

Another integral stage of our pipeline is to create RGBD frames from each pair of RGB and depth images, as shown in Fig. 1. We use the clean filtered depth image, corresponding undistorted RGB image and the necessary calibration (intrinsic as well as extrinsic) parameters to map the depth pixels with the color pixels. For each image pair, we project the depth pixels (x_d, y_d) to the 3D world space using Eqs. 7a, 7b, 7c, and then reproject them to the 2D color space and get the color using Eq. 8a–c, as in [6].

$$P3D.x = (x\_d - cx\_d) * depth(x\_d, y\_d)/fx\_d, \tag{7a}$$

$$P3D.y = (y\_d - cy\_d) * depth(x\_d, y\_d)/(fy\_d) \tag{7b}$$

$$P3D.z = depth(x\_d, y\_d), \tag{7c}$$

where fx_d, fy_d, cx_d, cy_d are the intrinsic parameters of the color sensor, and R and t are the estimated extrinsic parameters.

$$P3D' = R.P3D + t, \tag{8a}$$

$$P_2D_c.x = \left(P_3D'.x + \frac{fx\_c}{P_3D'.z}\right) + cx_c, \tag{8b}$$

$$P_2D_c.y = \left(P_3D'.y + \frac{fy\_c}{P_3D'.z}\right) + cy_c, \tag{8c}$$

where fx_c, fy_c, cx_c, cy_c are the intrinsic parameters of the color sensor, and R and t are the estimated extrinsic parameters.

One major challenge faced in this mapping is the resolution difference between the sensors. Consequently, when we map the depth pixels to color frame, we see a lot of missing pixel values. A morphological closing operation is performed to fill up those gaps.

3.4 Point Cloud Registration for Final Construction

In this section, we give a detailed description of how we align a pair of 3D point clouds (extracted from consecutive RGBD image frames) in order to achieve our final goal of creating accurate 3D models of dormant apple trees. Our registration framework is a three-stage process—first, the initial transformation is obtained from a consecutive pair of RGBD image frames, the second stage of transformation is obtained by aligning the corresponding pair of entire 3D point clouds, and finally, a fine tuning of alignment is performed on the pairs of semicircle fitted models whose 3D point clouds are found to be similar. The process of finding similar point cloud pairs is explained in Sect. 3.4.2.

3.4.1 Alignment of 3D Point Cloud Pair

In our work, we have used a combination of the basic Iterative Closest Point (ICP) Algorithm [31] and a modified version of it, known as RGBD-ICP [12, 14], to register a pair of 3D point clouds accurately. To obtain the initial alignment, we perform RGBD-ICP on a pair of consecutive RGBD images, as shown in Fig. 2a. Another stage of rigid transformation computation is achieved via a second run of basic ICP refinement on 3D point cloud pair obtained from the RGBD images.

RGBD-ICP

RGBD alignment framework involves extraction and matching of feature points from consecutive pair of RGBD images using a Scale Invariant Feature Transform (SIFT), and RANSAC based algorithm, known as RGB-D ICP, as in [12, 14]. In particular, the process starts with extracting sparse visual features from two consecutive frames, and then, associating them with their corresponding depth values to generate features points in 3D. RANSAC alignment, used as a line fitting technique, is performed between the matched feature sets to find best rigid transformation and finally, ICP refinement is performed. As ICP only works when both the point clouds are already closely aligned, RANSAC is used as an initial transformation tool to compensate for the fast motion between consecutive images, captured with a relatively noisy sensor.

3.4.2 3D Reconstruction Based on a Semicircle Fitted Algorithm:

In Sect. 3.2, we have discussed how a semicircle fitted model can be created from a single depth image. In this part, we describe our technique to merge all the 3D point clouds once they are aligned, and how to use information of the corresponding semicircle fitted models for that purpose.

We start by converting the depth images into RGBD data frames. Once, 3D point clouds are extracted from them, we align the first point cloud pair $(V_i, V_j)$$ in sequence, following the method stated in Sect. 3.4.1. We map the first point cloud

(V_i) to the frame of the second (V_j), and measure the similarity between them based on Jaccard similarity metric, as in [7]. Jaccard provides us with a range within which the point clouds are considered to be similar, and anything outside that range is discarded due to dissimilarity. If the pair is dissimilar, we move on to the next point cloud (V_{j+1}) in sequence.

On the other hand, if a pair of aligned point clouds is similar, we fit a semicircle model (ψ_i, ψ_j) of the point cloud pair, perform ICP refinement between them (as shown in Fig. 2c), and compute their average model ψ_{global}. Then, we align V_j with the next point cloud in sequence, V_{j+1}, in an incremental manner, whenever similarity criteria are satisfied. In particular, we measure the similarity between the transformed V_j and V_{j+1}, map ψ_{global} to the frame of V_{j+1}, perform registration refinement, compute the average of transformed ψ_{global} and ψ_{j+1}, and update the previous average model. We repeat this incremental process until all the images in the sequence are visited. At the end, we obtain one globally consistent semicircle fit model whose frame of reference coincides with that of the last image frame in our sequence. The incremental approach is described in Algorithm 1.

Algorithm 1 Creating a Globally Consistent 3D Model

Step 1: Initialization

1) Set $i = 1$. $j = 2$; where i and j are the indices for the previous and current point clouds as well as RGBD images

2) Initialize $\psi_{global} = \psi_1$. ψ_{global}, the final 3D model, is initialized to the semicircle fitted model of the first image

Step 2: Basic Operations

1) Convert X_i and X_j to V i and V_j. X_i and X_j and represent the 6D RGBD images. V i and V_j are the corresponding point clouds

2) Transform V i to the frame of V_j, as shown in Fig. 2(a) and Fig. 2(b)

3) Measure similarity between transformed V_i and V_j

4) If similar = TRUE, build the model ψ_j, and go to Step 3

5) If similar = FALSE, discard X_j increment j

Step 3: Creating Global Model

1) Transform ψ_{global} to the frame of V_j

2) Save ψ_{global} = Transformed ψ_{global}

3) Perform ICP refinement on ψ_{global} and ψ_j

4) Update ψ_{global} = average (ψ_{global}, ψ_j). Increment i, j

5) Repeat Steps 2 & 3 until all the RGBD image frames in the sequence are visited

4 Experimental Setup

In Sect. 4.1, we first provide the information of the dormant apple tree dataset that we have used to simulate the proposed 3D reconstruction technique. Then we describe the benchmark metrics used for evaluating the performance of our proposed framework in Sect. 4.2.

4.1 Data Sets

We have evaluated our algorithm on 5 different trees of the dataset[1] collected from several orchards of Indiana and Pennsylvania. The dataset contains data for 9 different dormant apple trees—three of these are present indoor, and rest are outdoor trees. Detailed description of the dataset can be found in [3].

4.2 Evaluation Metrics

To gain a thorough understanding of the quantitative performance of our proposed approach, three different evaluation metrics, namely, Branch Identification Accuracy (BIA), Estimation Error and Confidence Value are used. Detailed description of these metrics can be found in [7, 17]. Here, we would like to mention that the measure BIA is calculated based on the number of visible branches in the input data sequence of our algorithm. The confidence values and mean estimation errors are also computed based on the number of reconstructed branches.

5 Results and Analysis

In this section, we provide the results of our proposed 3D reconstruction approach explained in Sect. 3. Our algorithm has been simulated on a sequence of color and depth image pairs for 5 different trees present in the database [3], and the final reconstruction results have been displayed in Fig. 3. Table 1 presents a quantitative comparison of our results with a baseline approach [7] that utilizes only depth imaging information for reconstruction, and in Table 2, we compare the runtime complexities of our proposed algorithm and the baseline approach.

Tree 1 in Table 1 is a simple tree from Indiana, currently kept in indoor set-up. Our proposed approach achieves a mean estimation error of 1.84 mm, a remarkable improvement over 3.6 mm error of the baseline approach [7]. However, both modeling approaches reconstruct all the primary branches present in the tree achieving 100% BIA.

On the other hand, for **Tree 3**, an indoor tree from Pennsylvania, which has a very complex structure with thin and overlapping branches (as seen in Fig. 3b), only 10 out of its 11 branches could be reconstructed employing both the approaches producing a BIA of 90.9%. However, the mean estimation error of 2.26 mm for our approach is better than the 3.24 mm error of the depth reconstruction approach.

We have also applied our RGBD algorithm on **Tree 5**, an outdoor tree from Pennsylvania. All the data for this particular tree is collected in bright sunny weather (displayed in Fig. 3c) with the sensor facing sun directly. Our proposed approach

[1] https://engineering.purdue.edu/RVL/CVPRW_Dataset/

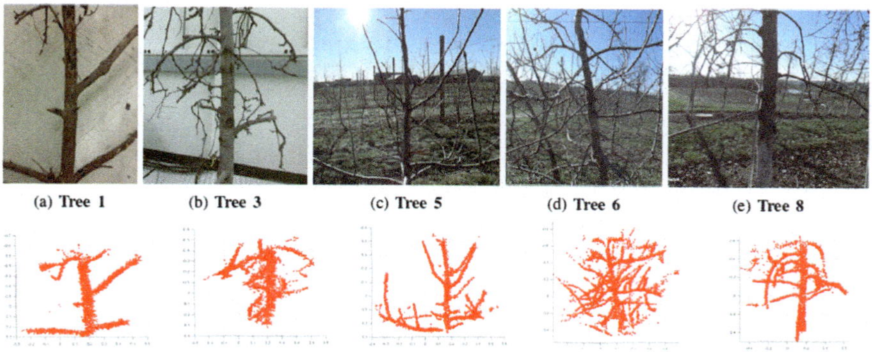

Fig. 3 Qualitative 3D Reconstruction Results for five apple trees. The first row shows the ground truth images, and the second row displays their corresponding 3D models using our proposed RGBD technique

gives 100% BIA in comparison to 83.3% BIA of the baseline technique. Here also, our proposed method outperforms the baseline approach with a mean estimation error of 2.12 mm over 3.58 mm.

For **Tree 6**, the mean estimation error using our approach is 1.87 mm better than 3.3 mm error of the depth based approach. Owing to its complex structure, we could reconstruct only 83.3% of the primary branches. However, this result also outperforms the baseline BIA of 72.2%.

Finally, we analyze our algorithm on another outdoor tree, **Tree 8**, from Pennsylvania. Our proposed approach gives 100% BIA in comparison to 88.9% BIA of the baseline technique, and also outperforms the baseline approach in terms of mean estimation errors and confidence values.

From Fig. 4a–b and Table 1, it is evident that our current technique shows significant improvement over the baseline approach in terms of BIA, confidence values and mean estimation errors. On an average, we could accurately estimate the diameters of 74% of primary branches within a tolerance of 3 mm, 92.2% of primary branches within an 5 mm error, and 100% of the primary branches within an error margin of 7 mm. We could identify the branches of the trees with 94.8% accuracy on an average, and that the mean estimation errors overall reduced when color information issued in addition to depth data for 3D reconstruction, as displayed in last row of Table 1.

In Table 2 and Fig. 4c, we show that our proposed algorithm also achieves a significant improvement over the depth reconstruction method in terms of the runtime complexity. We observe that the runtime depends on the tree structure as well. For example, **Tree 1** owing to its simple structure takes less time to run even for more number of frames. Similarly, **Tree 8** being more complicated in structure takes more time than **Tree 1** for the same number of image frames. Among all the trees, **Tree 6** takes largest time to be evaluated due to its extremely complicated structure.

Table 1 Quantitative comparison of our proposed RGBD reconstruction technique with a baseline approach using only depth information

(Tree ID, NI)	RGBD reconstruction					Depth reconstruction				
	Confidence values (mm)			Mean estimation error (mm)	BIA (%)	Confidence values (mm)			Mean estimation error (mm)	BIA (%)
	$\varepsilon = 3$ mm	$\varepsilon = 5$ mm	$\varepsilon = 7$ mm			$\varepsilon = 3$ mm	$\varepsilon = 5$ mm	$\varepsilon = 7$ mm		
(Tree 1, 30)	66.7	88.9	100.0	1.84	100.0	55.5	77.8	77.8	3.60	100.0
(Tree 3, 40)	80.0	90.0	100.0	2.26	90.9	60.0	80.0	90.0	3.24	90.9
(Tree 5, 25)	83.3	100.0	100.0	2.12	100.0	40.0	80.0	100.0	3.58	83.3
(Tree 6, 30)	73.3	93.3	100.0	1.87	83.3	53.8	92.3	92.3	3.30	72.2
(Tree 8, 30)	66.7	88.9	100.0	2.10	100.0	50.0	87.5	100.0	2.96	88.9
Mean	74.0	92.2	100.0	2.04	94.8	51.9	83.5	92.0	3.34	87.1

Table 2 Comparison of time complexity of our proposed RGBD approach with the baseline method

(Tree ID, NI)	Runtime for RGBD algorithm (in min)	Runtime for depth algorithm (in min)
(Tree 1, 30)	8.19	41.72
(Tree 3, 40)	17.45	63.76
(Tree 5, 25)	9.33	42.18
(Tree 6, 30)	26.89	137.80
(Tree 8, 30)	12.74	48.32

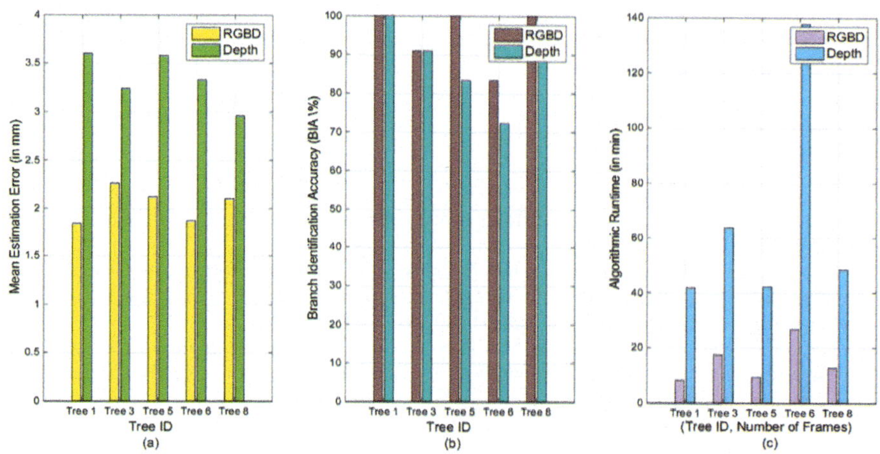

Fig. 4 Comparison of our proposed RGBD technique with the baseline approach in terms of **a** Mean estimation errors, **b** Branch identification accuracy (BIA), and **c** runtime complexity

6 Conclusion

In this chapter, we present a method that assimilates RGB information with time-of-flight depth data obtained using a Kinect2 sensor to create accurate 3D models of the trunk and primary branches of dormant apple trees. The main goal was to achieve accurate 3D reconstruction in absence of quality depth data and when the data is collected in outdoor environment in variable lighting condition. Both qualitative and quantitative evaluations have been performed by applying our proposed algorithm on several dormant apple trees under various indoor and outdoor environments, and the results look promising. Our experiments show that the proposed method gives a better estimation of the diameters and better runtime complexity when compared with the baseline approach that utilizes only time-of-flight depth information.

As mentioned previously, results cited in [2, 7] rely heavily on the quality of the depth images obtained from the sensor. As a step towards robustness, we started off with using RGBD information only for the initial alignment purpose. Though the improvement over the depth reconstruction approach was noticeable, we aimed for even more accuracy and thus, incorporated color information in the entire pipeline of our technique—alignment, 3D reconstruction and modeling. On an average, we could achieve a mean estimation error of 2.08 mm using our complete RGBD framework, in comparison to the best mean estimation error of 2.36 mm utilizing RGBD information only for alignment purpose.

In the future, our plan is to include a color constancy model in our pipeline for compensating the effect of direct sun on the RGBD sensor while capturing images. We would also focus on developing a more robust skeletonization algorithm from 2D range images and, invest our time towards developing a novel calibration algorithm for RGBD sensors. Finally, we plan to further improve the computation time of our algorithm via GPU computation and parallelization.

References

1. B. Adhikari, M. Karee, 3d reconstruction of apple trees for mechanical pruning. Am. Soc. Agricult. Biol. Eng. (2012)
2. S. Akbar, N. Elfiky, A. Kak, A novel framework for modeling dormant apple trees using single depth image for robotic pruning application. In Proceedings of the IEEE International Conference on Robotics and Automation (ICRA), (2016)
3. S. A. Akbar, S. Chattopadhyay, N. M. Elfiky, A. Kak, A novel benchmark RGBD dataset for dormant apple trees and its application to automatic pruning. In The IEEE Conference on Computer Vision and Pattern Recognition (CVPR) Workshops, (2016)
4. P.J. Besl, N.D. McKay, Method for registration of 3d shapes. In Robotics—DL tentative. International Society for Optics and Photonics, (1992), pp. 586–606
5. M. Brophy, A. Chaudhury, S.S. Beauchemin, J.L. Barron, A method for global non-rigid registration of multiple thin structures. In Computer and Robot Vision (CRV) (2015), pp. 214–221
6. N. Burrus, Kinect calibration. 2014. http://nicolas:burrus:name/index:php/Research/KinectCalibration
7. S. Chattopadhyay, S.A. Akbar, N.M. Elfiky, H. Medeiros, A. Kak, Measuring and modeling apple trees using time-of-flight data for automation of dormant pruning applications. IEEE Winter Conference on Applications of Computer Vision (WACV) (2016) , pp. 1–9
8. K.N. Chaudhury, Y. Khoo, A. Singer, Global registration of multiple point clouds using semidefinite programming. SIAM J. Optim. **25**(1), 468–501 (2015)
9. Y. Cui, S. Schuon, S. Thrun, D. Stricker, C. Theobalt, Algorithms for 3d shape scanning with a depth camera. Patt. Anal. Mach. Intell. IEEE Trans. **35**(5), 1039–1050 (2013)
10. N. Elfiky, S. Akbar, J. Sun, J. Park, A. Kak, Automation of dormant pruning in specialty crop production: an adaptive framework for automatic reconstruction and modeling of apple trees. In Proceedings of the IEEE Conference on Computer Vision and Pattern Recognition Workshops (2015), pp. 65–73
11. I. Hameduddin, Rotate vector(s) about axis., (2012)
12. P. Henry, M. Krainin, E. Herbst, X. Ren, D. Fox, Rgbd mapping: using depth cameras for dense 3d modeling of indoor environments. In The 12th International Symposium on Experimental Robotics (ISER) (2010)
13. P. Henry, M. Krainin, E. Herbst, X. Ren, D. Fox, Rgbd mapping: using kinect-style depth cameras for dense 3d modeling of indoor environments. Int. J. Robot. Res. **31**(5), 647–663 (2012)
14. P. Henry, M. Krainin, E. Herbst, X. Ren, D. Fox, Rgbd mapping: using depth cameras for dense 3d modeling of indoor environments. In Experimental robotics (2014), pp. 477–491

15. S. Izadi, D. Kim, O. Hilliges, D. Molyneaux, R. Newcombe, P. Kohli, J. Shotton, S. Hodges, D. Freeman, A. Davison, Kinectfusion:real-time 3d reconstruction and interaction using a moving depth camera. In Proceedings of the 24th annual ACM symposium on User interface software and technology (2011), pp. 559–568
16. S. Jin, L. Fan, Q. Liu, R. Lu, Novel calibration and lens distortion correction of 3d reconstruction systems. J. Phys.: Conf. Series **48**, 359 (2006)
17. M. Karkee, B. Adhikari, S. Amatya, Q. Zhang, Identification of pruning branches in tall spindle apple trees for automated pruning. Comput. Electron. Agric. **103**, 127–135 (2014)
18. Y. Livny, F. Yan, M. Olson, B. Chen, H. Zhang, J. ElSana, Automatic reconstruction of tree skeletal structures from point clouds. ACM Trans. Graph. (TOG) **29**, 151 (2010)
19. L.D. Lopez, Y. Ding, J. Yu, Modeling complex unfoliaged trees from a sparse set of images. Comput. Graph. Forum **29**, 2075–2082 (2010)
20. T. Mallick, P.P. Das, A.K. Majumdar, Characterizations of noise in kinect depth images: a review. Sens. J. IEEE **14**(6), 1731–1740 (2014)
21. S. Meister, S. Izadi, P. Kohli, M. Hammerle, C. Rother, D. Kondermann, When can we use kinectfusion for ground truth acquisition? Proc. Workshop on Color-Depth Camera Fusion Robot., **2** (2012)
22. C.V. Nguyen, S. Izadi, D. Lovell, Modeling kinect sensor noise for improved 3d reconstruction and tracking. In 3D Imaging, Modeling, Processing, Visualization and Transmission (3DIMPVT) (2012), pp. 524–530
23. I. Shlyakhter, M. Rozenoer, J. Dorsey, S. Teller, Reconstructing 3d tree models from instrumented photographs. IEEE Comput. Graphics Appl. **3**, 53–61 (2001)
24. Y.B. Sinai, Skeletonization using voronoi (2010)
25. J. Smisek, M. Jancosek, T. Pajdla, 3d with kinect. In Consumer Depth Cameras for Computer Vision (2013), pp. 3–25
26. A. Tabb, Three-dimensional reconstruction of fruit trees by a shape from silhouette method. Am. Soc. Agricult. Biol. Eng., 1 (2009)
27. G.K. Tam, Z.-Q. Cheng, Y.-K. Lai, F.C. Langbein, Y. Liu, R.R.M.D. Marshall, X.-F. Sun, P.L. Rosin, Registration of 3d point clouds and meshes: a survey from rigid to nonrigid. Vis. Comput. Graph. **19**(7), 1199–1217 (2013)
28. A. Velizhev, Gml c++ camera calibration toolbox. consulted on, **19**(05) (2015)
29. Q. Wang, Q. Zhang, Three-dimensional reconstruction of a dormant tree using rgb-d cameras. Am. Soc. Agricult. Biol. Eng., (2013)
30. T. Weber, R. Hansch, O. Hellwich, Automatic registration of unordered point clouds acquired by kinect sensors using an overlap heuristic. ISPRS J. Photogramm. Remote. Sens. **102**, 96–109 (2015)
31. J. Wilm, Iterative closest point (2013)
32. N. Elfiky, Application of analytics in machine vision using big data. Asian J. Appl. Sci. **7**(4), 376–385 (2019)
33. D.M. Kirkpatrick, K.B. Rice, A. Ibrahim, S.J. Fleische, The influence of marking methods on mobility, survivorship, and field recovery of Halyomorpha halys (Hemiptera: Pentatomidae) adults and nymphs. Environmental entomology (2020)
34. M.J. Feldmann, A. Tabb, Cost-effective, high-throughput phenotyping system for 3D reconstruction of fruit form. Plant Phenom. J., (2022)

Wheat Leaf-Disease Detection Using Machine Learning Techniques for Sustainable Food Quality

Rania El-Sayed, Ashraf Darwish, and Aboul Ella Hassanien

Abstract Wheat is considered one of the best food resources and quality in many countries, especially in Africa and Asia. However, the crops are affected by many diseases during the cultivation process including rust and leaf spots that can destroy the crop after a month. This leads to a large degradation in wheat productivity rate. So, the crops need continuous observation during the agriculture process and need experts with the wheat leaf's nature. This is a time-consuming and expensive task. So, this paper aims to detect various types of leaf diseases to differentiate them from healthy crops without needing continuous crop monitoring and to minimize yield losses. The dataset used is a three-class wheat leaf dataset that consists of one healthy class and 2 leaf disease classes namely septoria and stripe rust. We proposed a Support Vector Machine (SVM) DL-based features model that depends on features extracted from different deep Transfer Learning (TL) models applied to the dataset after pre-processing that are then passed to the SVM classifier. VGG16, VGG19, and InceptionResNetV2 models are used to extract the features out of the images one at a time, where, VGG19 is the best model where it achieves an accuracy of 98% which outperforms many researches in the literature.

Keywords Deep learning · Transfer learning · VGG19 · SVM · Wheat leaf-disease · Rust

R. El-Sayed
Computer Engineering, Helwan University, Cairo, Egypt

A. Darwish
Faculty of Science, Helwan University, Cairo, Egypt

A. E. Hassanien (✉)
Faculty of Computers and Artificial Intelligence, Cairo University, Cairo, Egypt
e-mail: aboitcairo@cu.edu.eg

A. Darwish · A. E. Hassanien
Scientifc Research Group in Egypt (SRGE), Cairo, Egypt

A. E. Hassanien and M. Soliman (eds.), *Artificial Intelligence: A Real Opportunity in the Food Industry*, Studies in Computational Intelligence 1000,
https://doi.org/10.1007/978-3-031-13702-0_2

1 Introduction

Agriculture is the most crucial sector of the economy for most countries. However, the overall crop yield is low since most crops are subjected to many diseases. One of the most common diseases that affect crops is the leaf disease including rust, gray leaf spot, and leaf blight which are common for some crops like rice, wheat, corn, and many others. Since leaf disease is more common, it will be the scope of this paper. Crop leaf disease can be detected through visual observation, however, this requires a team of specialists and constant crop monitoring which is a highly expensive, and time-consuming task. For these reasons, the visual observation technique is less reliable. So, Deep Learning (DL) is involved to solve such problems in a more reliable, fast, accurate, and precise way.

DL is an end-to-end approach that is highly required for various classification tasks due to its high performance in image classification and analysis. However, Transfer Learning (TL) is widely used nowadays for these tasks since the training time can be reduced compared to the time needed by a DL model built from scratch. That is because TL models are pre-trained, so only the last fully connected layers are eliminated and replaced with other dense layers or fine-tuned to match the specific task of leaf disease classification. These replaced layers are trained on the new task without needing to re-train the whole architecture.

Food insecurity is described as the inability to get sufficient and appropriate food, or the possibility that a person or a household will go without food. Food security is defined by the United Nations Food and Agriculture Organization (FAO) as having four dimensions: availability, access, use, and stability. Over the last 50 years, the definition of food security has expanded and altered, including the establishment of the four often referenced pillars of food security: availability, access, utilization, and stability, which have been critical in defining policy. Food quality governance may be defined as establishing circumstances for food quality standards that do not rely only on government power and sanction, but are the product of interactions between public, commercial, and volunteer sector players. In recent decades, there has also been a rising awareness of major concerns affecting hunger and malnutrition, such as increasing food system inequities typified by unequal power dynamics and deteriorating global climatic and ecological catastrophes.

So, in this paper, the main goal is to use the power of DL algorithms to classify the wheat leaf image into one of three classes healthy, septoria, and stripe rust. The latter two are two types of leaf diseases that affect wheat crops. We proposed an SVM DL-based feature model, where the input is the features extracted from each wheat leaf image using different DL models, and the output is one of the three classes of the wheat leaf included in the dataset as discussed in Sect. 3.2. Before the feature extraction, the dataset is resized and scaled. The TL models used for feature extraction in this paper are VGG16, VGG19, and InceptionResNEtV2. The main goal of this paper is to extract the features from each model and then compare their results after applying classification using the SVM classifier. The TL models are trained on a publicly available dataset with 407 images composing the three classes mentioned above. Due to the challenge of having a very small dataset, we perform some pre-processing steps on it, as discussed in Sect. 3.1.

The main contributions of this paper are:

- Discussing the concepts behind each leaf disease.
- Proposing an SVM DL-based features based on different TL models to compare which one is more efficient to differentiate between different leaf diseases from healthy leaves.
- Discussing the results with other models proposed in the literature with different datasets.
- To our best knowledge, our model achieves the highest performance with the used dataset among other models in the literature.

  The rest of the paper is organized as follows: Sect. 2 shows other related work. Section 3 discusses the proposed approach. Section 4 shows the experimental results. Section 5 concludes the paper and shows what can be done as future work.

2 Related Work

There are many researches in the literature that focuses on the agriculture problem of many crops such as tomato, cotton, rice, wheat, and many other. Some of these researches are listed below.

Poojan Panchal et al. [1] constructed a dataset that is composed of four classes including Bacterial Spot, Early Blight, and Late blight as leaf diseases and a healthy leaf class. Then the authors applied the HSV-Herpes Simplex Virus segmentation technique for better feature extraction, to partition the healthy and affected portions. Then they used a random forest classifier for plant leaves disease detection on the Gray Level Co-occurrence Matrix (GLCM) features extracted from the images and achieved an accuracy of 98%.

Jiang et al. [2] proposed a multi-task model that detects the type of leaf disease for rice and wheat crops in the same model using a shared layer, where every epoch is trained alternately. They used two publicly available datasets for rice and wheat including these classes: rice bacterial leaf blight, rice brown spot, rice leaf smut, wheat rust, and wheat powdery mildew. Then they applied some pre-processing steps like rescale, and resize before training the tuned VGG16 TL model. They achieved 97.22% and 98.75% for rice leaf diseases and wheat leaf disease respectively.

Krishnamoorthy N et al. [3] proposed a simple CNN and InceptionResNetV2 TL models for rice leaf disease detection. They used a publicly available dataset on Kaggle with a total of 5200 images composed of four classes: leaf blast, bacterial spot, bacterial blight, and healthy crop. The first three classes represent diseases, and the fourth class represents a healthy crop. They applied pre-processing steps like image augmentation, rescale, and resize. They achieved an accuracy of 84.75%

with 15 epochs and 95.67% with 10 epochs for simple CNN and InceptionResNetV2 models respectively.

Solemane Coulibalya et al. [4] added two fully connected layers over the pre-trained VGG16 model to detect the presence or absence of mildew disease for the millet crop. They applied pre-processing steps like data augmentation and image crop, where they achieved an accuracy of 89% with the VGG16 model which trained on two classes out of 124 leaf images of millet crop.

Nandhini et al. [5] proposed an algorithm to classify plant leave diseases. They applied multi-pre-processing steps on the images for better performance like image segmentation, then extracted features from segmented images using GLCM and applied SVM, KNN, and DTs on the extracted features as classifiers, where SVM achieved the highest performance of 98.51%.

Shreya Ghosal et al. [6] proposed a VGG16 model that is trained on their constructed dataset that is composed of four rice leaf disease classes, and they achieved a 92.46% accuracy.

3 Proposed Approach

3.1 Dataset

The dataset used in this paper is publicly available on the Kaggle [7]. It consists of three classes: Healthy, Septoria, and Stripe Rust. A sample for each class is shown in Fig. 1. The dataset consists of 102 images for healthy crops, 97 images for Septoria leaf disease, and 208 images for Stripe Rust leaf disease. Since the dataset is small and unbalanced, some pre-processing steps are needed, which are covered further in this section.

3.2 Approach

This section discusses the architecture of the proposed model, its parameters, and how it works in detail. As mentioned earlier in this paper, the main goal is to classify and detect the leaf disease of the wheat crop only. One of the main leaf diseases of the wheat crop is leaf rust where there are three types of rust [8]. The first type is the yellow or stripe rust, a fungal disease that can attack the wheat crop at any stage of the wheat life cycle, where it affects the weight, size, and the number of grains per spike for the wheat crop in cold temperatures at most. The second type is the brown or leaf rust, the most common wheat rust fungal disease, which reduces the number of grains per wheat and damages the leaf sheaths and glumes under favorable weather conditions. The last type is the black or stem rust, the most detrimental fungal disease for cereal crops in the world, where it affects stem, sheaths, glumes,

a b c

Fig. 1 A sample of each wheat class in the dataset. **a** Healthy wheat leaf, **b** Septoria leaf disease, and **c** Stripe Rust leaf disease

and spikes, reduces kernels per spike, and shrunk grain size in warm temperatures. The three types of rust are shown in Fig. 2. In this paper, the main focus is only on one type of rust, the stripe rust since it is the most destructive one. The other type of leaf disease this paper focuses on is the septoria, which is one of the second most important wheat diseases after rust that causes a significant threat to global food production. It causes degradation of wheat production by 35–50%. It is mostly active in cold temperatures.

a **b** **c**

Fig. 2 A sample of each leaf rust disease type of the wheat crop. **a** Yellow/Stripe rust, **b** Brown/Leaf rust, and **c** Black/Stem rust

As mentioned in Sect. 3.1, the dataset is small and is not balanced, so some pre-processing steps are applied to it before applying the training model. Among these pre-processing steps:

- Data resize: all the images are resized to 224×224.
- Data re-scale: all the images are re-scaled through dividing each pixel by 255 to force the values of pixels between 0 and 1.
- Data balancing: SMOTE technique is used to up-sample the data so that all classes have equal number of samples.
- Data HSI-segmentation: all the images are converted from BGR to HIS colors then thresholding is applied to segment each object in the image.

These pre-processing steps help having more samples, as well as, having all scenarios for captured images. A sample of each class after pre-processing is shown in Fig. 3.

In this paper, to solve the problem of leaf disease monitoring for the wheat crop, we used an SVM classifier trained on DL features extracted from different TL models. The features are extracted from the processed images using InceptionResNetV2, VGG16, and VGG19 TL models one at a time to compare which of them is more effective. To extract the features, the classification layer for each TL model is removed to have the flattened features as the final output. Then the extracted features are passed to the SVM classifier with a linear kernel to give the output as the disease type of the wheat leaf. The overall architecture of this model is shown in Fig. 4.

TL models are DL models that are pre-trained on the ImageNet dataset, a dataset that is composed of 1000 classes of different categories. Each TL model is composed of 2D convolutional, max-pooling, and dense or fully connected as the basic layers. Some other layers can be added for optimization such as Dropout and Batch Normalization layers. The last fully connected layers in any model are used for classification, so to use any of the TL models as a feature extractor, the last fully connected layers are removed since the goal is to just get the output features before the classification phase not to classify.

The VGG16 and VGG19 TL models are almost the same. Both are Convolutional Neural Network (CNN) architectures that are considered one of the best vision models. Both have the same input and output shapes. One difference is the number of convolutional layers, where VGG16 has 16 convolutional layers, while VGG19 has 19 convolutional layers. The other difference is the total number of parameters, where VGG19 has more parameters. The architectures of both models are shown in Figs. 5 and 6 respectively.

The InceptionResNetV2 [9] model is a combination of Inception and ResNet models. The inception network performs a convolution operation on the input with three different filter sizes: 1×1, 3×3, and 5×5, also a max-pooling operation is applied then the output of all operations is concatenated as shown in Fig. 7. For more efficient and fewer computations, an extra 1x1 convolution filter is added before the 3×3 and 5×5 convolutions and it is added after the max-pooling filter as shown in Fig. 8. This is done to reduce the number of input channels, hence needing fewer

Fig. 3 A sample of each wheat leaf disease after applying image pre-processing. The images **a**, **b**, and **c** are Healthy Wheat Leaf, Septoria Leaf Disease, and Stripe Leaf rust classes before applying HSI-Segmentation, while **d**, **e**, and **f** are Healthy Wheat Leaf, Septoria Leaf Disease, and Stripe Leaf rust classes after applying HSI-Segmentation

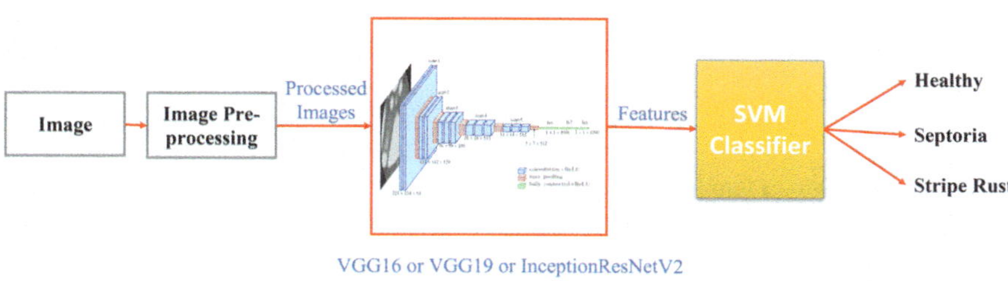

Fig. 4 The architecture of the SVM DL-based features model

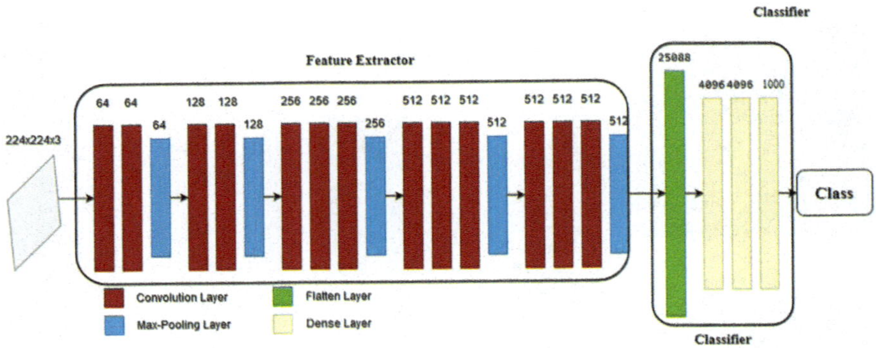

Fig. 5 The architecture of the VGG16 convolutional neural network model

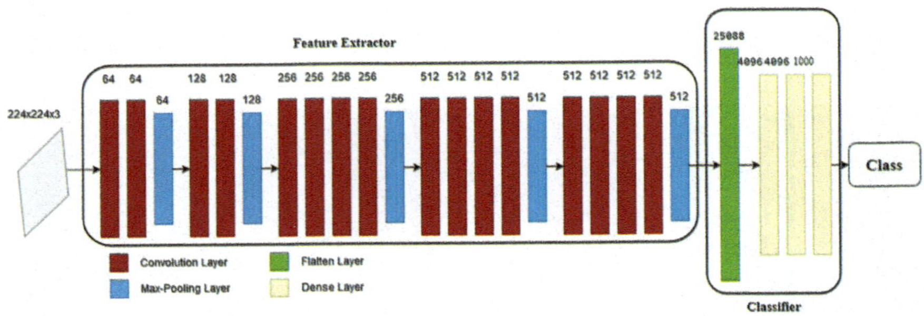

Fig. 6 The architecture of the VGG19 convolutional neural network model

computation resources. The InceptionV2 updates the Inception module by only splitting the 5×5 convolution filter into two 3×3 filters to improve the computational speed. The overall architecture includes nine inception modules and is 22 layers deep. The ResNet model concept is to skip some operations to reduce computations also as shown in Fig. 9. The goal behind adding the Inception module to the residual block is that the residual block adds the output of the convolution operation of the inception module, to the input as shown in Fig. 10.

Concerning the data balancing, SMOTE technique is used which stands for Synthetic Minority Oversampling Technique. It is an up-sampling technique that increases the number of samples in the minority class depending on the neighbors of the sample.

4 Experimental Results

This section shows and compares the results of the proposed model as well as discusses the results of other models in the literature. All experiments are conducted

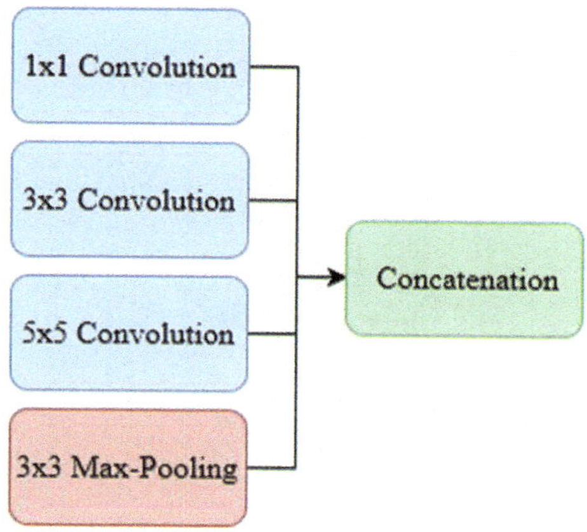

Fig. 7 The block of inception module

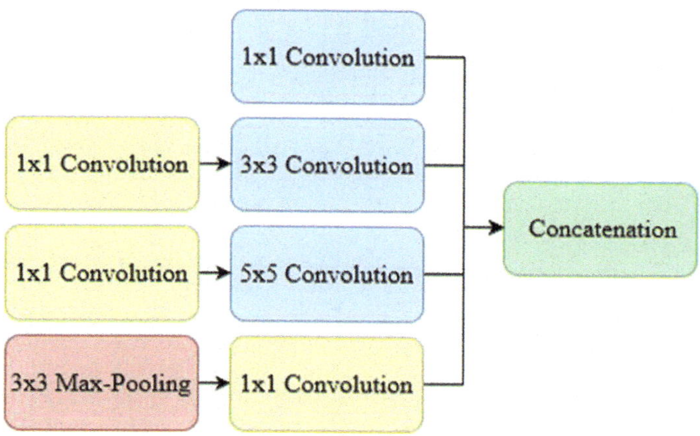

Fig. 8 The block of inception module with dimension reduction

on the GPU of the colab notebook. The dataset is divided into 70% train, 15% test, and 15% validation sets, where features are extracted for each set separately at a batch size equal to 1 and target image size equal to 224×224.

For our proposed model, three experiments are conducted, where, in each experiment, one TL model is used to extract the features from the HSI- segmented images. Then passing the features to the SVM classifier. The experiments include feature extracting through one of these models: VGG16, VGG19, and InceptionResNetV2. The number of the extracted features through each TL model is shown in Table 1.

Table 2 shows the results of applying the SVM classifier to the HSI- segmented images. As shown, the features extracted from VGG19 are the most efficient and achieve an accuracy of 98%. However, the features of the VGG16 model are not that

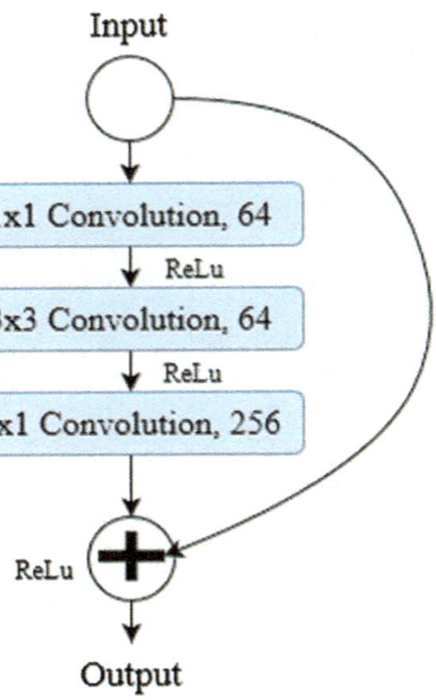

Fig. 9 The residual block of resnet model

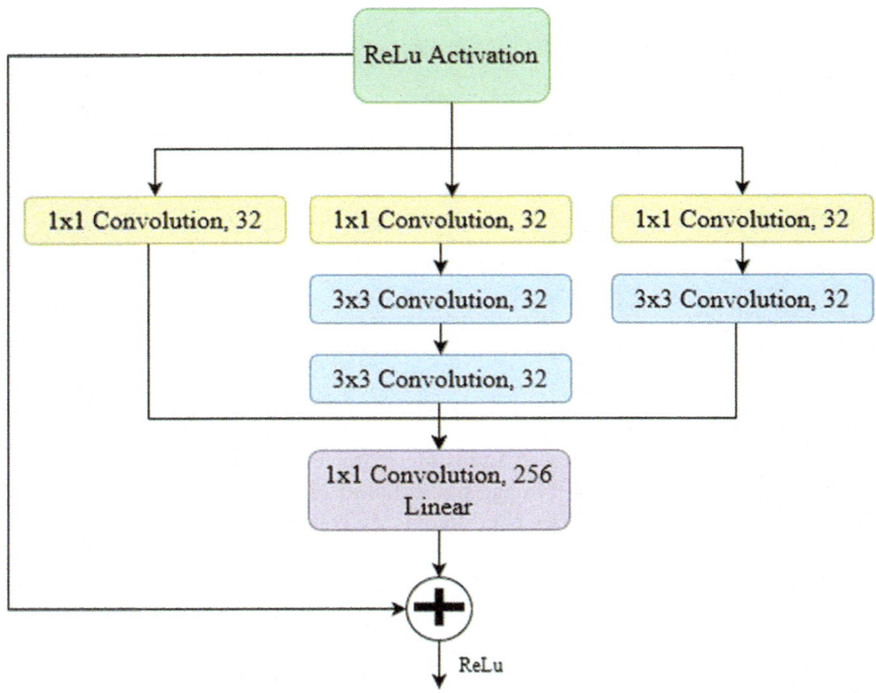

Fig. 10 The inceptionresnet block

Table 1 The number of the extracted features for each TL model

Model	Number of features (K)
VGG16	25.18
VGG19	25.18
InceptionResNetV2	38.4

Table 2 The results of SVM DL-based features model on HSI-segmented images

Feature extractor	Precision (%)	Recall (%)	F1-score (%)	Accuracy (%)
InceptionResNetV2	91	91	91	91
VGG16	97	94	95	95
VGG19	99	97	98	98

bad. The features of VGG16 may be better with more tuning or if gets enhanced features through further processing on input images. The InceptionResNetV2 has a smaller accuracy since the model is too complex for this small dataset. In the literature, Jiang et al. [2] proposed a multi-task model to detect the type of wheat and rice leaf diseases. For the wheat crop, they have two classes wheat rust and wheat powdery mildew and they achieved an accuracy of 98.75%. However, they used a different dataset and different pre-processing techniques. So, to the best of our knowledge, our proposed SVM DL-based features achieve the highest performance in the literature on the used wheat leaf disease dataset.

5 Conclusion and Future Work

Recent breakthroughs in human development theory are essential to the increased emphasis on food security. More sustainable food systems are within reach if supported by government policy and research. Food quality schemes are multifaceted concepts developed on the interplay of distinctive constituents that the EU has developed. The overall quality of a food product is the consequence of the interaction of several factors. Food scientists are employing modern and multivariate equipment to analyze food quality aspects and variations more than ever before. The implementation of these methodologies necessitates the employment of appropriate data analysis techniques in order to extract useful and conclusive information from these massive data sets.

In this paper, multiple experiments are conducted on the proposed model to classify wheat leaf crop diseases. The model is based on deep features extracted from different TL models that are passed to the SVM classifier where the input images passed to the feature extractors are HSI-Segmented images. The model is trained on a publicly available wheat leaf disease dataset with three classes. From the results, it is obvious that the VGG19 feature extractor has the highest performance with an accuracy of 98% that outperforms other models in the literature. For future work, this model may be enhanced to classify other types of wheat leaf diseases or other crops like rice, tomato, corn, apple, and cotton.

References

1. P. Panchal, V.C. Raman, S. Mantri, Plant diseases detection and classification using machine learning models. In 2019 4th International Conference on Computational Systems and Information Technology for Sustainable Solution (CSITSS), vol 4 (IEEE, 2019), pp. 1–6
2. Z. Jiang, Z. Dong, W. Jiang, Y. Yang, Recognition of rice leaf diseases and wheat leaf diseases based on multi-task deep transfer learning. Comput. Electron. Agric. **186**, 106184 (2021)
3. D. Krishnamoorthy, V.L. Parameswari, Rice leaf disease detection via deep neural networks with transfer learning for early identification. Turkish J. Physiotherapy Rehabil., **32**, 2
4. S. Coulibaly, B. Kamsu-Foguem, D. Kamissoko, D. Traore, Deep neural networks with transfer learning in millet crop images. Comput. Ind. **108**, 115–120 (2019)
5. S. Aasha Nandhini, R. Hemalatha, S. Radha, K. Indumathi, Web enabled plant disease detection system for agricultural applications using wmsn. Wireless Pers. Commun. **102**(2), 725–740 (2018)
6. S. Ghosal, K. Sarkar, Rice leaf diseases classification using cnn with transfer learning. In 2020 IEEE Calcutta Conference (CALCON), (IEEE, 2020), pp. 230–236
7. Wheat Leaf Disease Dataset. https://www.kaggle.com/datasets/olyadgetch/wheat-leaf-dataset
8. U. Shafi, R. Mumtaz, Z. Shafaq, S.M.H. Zaidi, M.O. Kaifi, Z. Mahmood, S.A.R. Zaidi, Wheat rust disease detection techniques: a technical perspective. J. Plant Dis. Prot., 1–16 (2022)
9. InceptionResNetV2 Documentation. https://towardsdatascience.com/a-simple-guide-to-the-versions-of-the-inception-network-7fc52b863202

Explainable AI and Slime Mould Algorithm for Classification of Pistachio Species

Gehad Ismail Sayed and Aboul Ella Hassanien

Abstract The safety and quality of the food are considered an essential issue in the entire world. This is due to food being the basis of human health. Nowadays, machine learning algorithms have embodied the recent technology in all stages of food processing such as food grading, food quality determination, and food classification. Pistachio nuts have an important role in the agricultural economy. To increase the efficiency of post-harvest industrial processes, there is a need to introduce technologies for classifying different species of pistachio. This study considers an automated model to separate pistachio species. The proposed pistachio species classification consists of three main phases; features selection based on slime mould algorithm phase, feature interpretation based on explainable artificial intelligence phase, and finally classification of pistachio species using logistic regression phase. The proposed pistachio species classification model obtained overall 90% classification accuracy, 90% precision, and 91% f1-score.

Keywords Explainable artificial intelligence · Swarm intelligence · Feature selection · Pistachio species · Food Industry

1 Introduction

There are many factors associated with the processing of food products. These factors can result in food processing failing. Food processing has a great impact on economic development. There are many different aspects of food processing such as manufacturing and modification of food [5]. To satisfy high standards and expectations in

G. I. Sayed (✉)
School of Computer Science, Canadian International College (CIC), New Cairo, Egypt
e-mail: Gehad_Sayed@cic-cairo.com

A. E. Hassanien
Faculty of Computers and Artificial Intelligence, Cairo University, Giza, Egypt

G. I. Sayed · A. E. Hassanien
Scientific Research Group in Egypt (SRGE), Cairo, Egypt

A. E. Hassanien and M. Soliman (eds.), *Artificial Intelligence: A Real Opportunity in the Food Industry*, Studies in Computational Intelligence 1000,
https://doi.org/10.1007/978-3-031-13702-0_3

the food industry, the quality control of agriculture products must be efficient and accurate [13]. Over the years, the demand for food has increased with the increasing population, especially in developing countries such as Egypt, India, and China. Machine learning algorithms have proven their efficiency and effectiveness in diverse sectors such as medical care, drug discovery, industry, and transportation.

The authors in [6] used multilayer perceptron (MLP) and adaptive network-based fuzzy inference system (ANFIS) for the prediction of food production. They applied their model to agricultural productions data collected in Iran from 1961 to 2017. The results showed that their model can provide a suitable tool for policymakers in the prediction of the future of food production. In [11], authors used support vector machine (SVM) and Partial Least Squares Regression (PLS) for automatic classification of raw food. The experimental results showed that their proposed model can efficiently classify 7 different types of raw food. Another prediction model proposed in [1], where multiverse optimization algorithm (MOA) is applied to train ANFIS and MLP. The authors used the Bayesian average model to predict tomato yield crop. The experimental results revealed that their proposed model can be useful for food security and agriculture management.

Pistachio is considered one of the most nutritious food. Each 100 gr of pistachio provides 560 calories. Additionally, it has a substantial source of dietary fiber, protein, vitamins, and different dietary minerals. Pistachio species have many other benefits that can positively enhance heart health [7]. The most commonly used species are Siirt and Kirmizi since both of them have high economic value. In the pastry industry and sweet confectionery, Kirmizi pistachio species are frequently used due to their taste, dark green color, and distinct aroma. While Siirt pistachio species are used as snacks between meals due to their round shape and their high cracking rate. Machine learning algorithms can provide a great tool for classifying multi-categories of pistachios. Also, they can provide a fast classification speed compared to the traditional classification based on the human eyes. Machine vision-based algorithms calculate multiple features from each image. These features are used to feed a machine-learning algorithm to predict which type is this pistachio image. Authors in [7] used Otsu's method to segment each pistachio image into two segments one represents the pistachio region and the other segment is the background region. Then shape and morphological features are extracted from the pistachio region, then these features are used to feed K-NN classifier. The results showed that their proposed model obtained a classification success of 94.18%. Authors in [2] used the patches extraction method to extract a set of features from Jumbo, Long, Round, and Super long pistachios. They used EfficientNet-B1 convolutional neural network for pistachio classification. Their proposed model obtained overall 97.2% accuracy.

This study introduces a new pistachio species classification model, where slime mould algorithm and explainable artificial intelligence along with logistic regression classifier are used. Slime mould algorithm (SMA) is one of the recent swarm intelligence algorithms proposed in 2020 by Li et al. [4]. The main inspiration of SMA came from the oscillation behavior of slime mould in nature. This study considers SMA to find an appropriate feature subset. Selecting the optimal feature subset can significantly improve a classier's performance and decrease the computational time

[9, 10]. The selected features of SMA based feature selection algorithm are evaluated and interpreted using the Local Interpretable Model-Agnostic Explanations (LIME) method.

The organization of the rest of the chapter is structured as follows. A brief description of the basic SMA algorithm with a brief description of explainable artificial intelligence is introduced in Sect. 2. In Sect. 3, the proposed pistachio species classification model is described in detail. Section 4 describes the adopted dataset. Section 5 provides the simulation results and discussions. Finally, conclusions and future work are proposed in Sect. 6.

2 Basics and Background

2.1 *Explainable Artificial Intelligence*

Explainable Artificial Intelligence or shortly (XAI) allows us to interpret and explain the prediction results of machine learning algorithms [12]. Usually, XAI refers to techniques and methods which used in artificial intelligence applications. As the results of XAI techniques and methods can be understood by a human. Due to the popularity of complex machine learning algorithms, there are always existed a trade-off between the algorithm's performance and its explainability. XAI methods explain the prediction results of a machine learning algorithm.

Most complex machine learning algorithms face "Black-Box" implementation, which means that there is no simplified format to explain the prediction results. XAI methods are not mainly designed to help only data scientists but also designed to debug a machine learning model. Thus XAI methods can apply any machine learning algorithm in a white-box approach, where data scientists can effectively analyze their model and infer and understand if their model made the right prediction on the test samples. Therefore any discrepancy in the machine learning model can be detected and fixed through the development stage.

Local Interpretable Model-Agnostic Explanations (LIME) is one XAI method [8]. The main objective of LIME is to find an interpretable model over an interpretable representation that is locally faithful to a supervised machine learning algorithm. In other words, this method is used to explain a machine learning classifier for an instance. Thus LIME is suitable for local explanations. The interpretable model means that this model is used to explain a machine learning model that is s understandable to humans despite its actual features. For example, in-text classification, the interpretable representation can be a binary vector referring to the absence or the presence of a single word. Similar to the image classification task, it can be a binary vector as well indicating the absence or the presence of a contiguous patch of similar pixels. LIME starts with manipulating the input instance, then creates a sequence of artificial samples having only a part of the original attributes. The main privilege of the LIME explainable AI method is that it is compatible with many machine learning classifier algorithms. Additionally, it can be used with image, text, and tabular data. Let G denotes a class of potential models such as decision tree, logistic regression, and support vector machine. The explanation of the presentence or the absence for a model can be defined as $g \in G$ with a search boundary equals to $0, 1^d$, where zero denotes the absence of the interpretable component and one denotes the presence of the interpretable component. Let f denotes the machine learning model being explained and $f(x)$ is the probability that an instance x belongs to a predefined class. $\pi - x(z)$, where z is an instance, and $\pi - x(z)$ is the proximity measure from z to x. $\zeta(f, g, \pi - x)$ is used to indicate how unfaithful g is in the locality $\pi - x$ for approximating f. Finally, let $\Omega(g)$ is the measure of complexity such as the depth of tree in the decision tree algorithm. The overall explantation obtained from LIME is defined in Eq. (1), where this formula can be used to explain different algorithms G.

$$\Im(x) = \text{argmin}\zeta(f, g, \pi - x) + \Omega(g), \ \forall g \in G \tag{1}$$

2.2 *Slime Mould Algorithm (SMA)*

Inspiration Analysis Slime Mould Algorithm (SMA) is one stochastic-based optimization algorithm published in 2020 by Li et al. [4]. SMA inspired the oscillation behavior of slime mould in nature. The original SMA simulates the morphological attitude of slime mould during the foraging. Slime mould is one of the eukaryotes that survive in cold and humid places. Plasmodium is the essential nutritious phase. While the active and dynamic phase of slime mould is considered as the main research phase of developing the original SMA. During this phase, slime mould searches and surrounds the foods then conceals enzymes to digest. Due to the characteristics of slime mold patterns, slime mould can use multiple food sources to create an intravenous link combining them. Slime mould can filter agar and oatmeal. Thus it can be further used as prototypical creatures. Slime mould can make the best food path to link food in a relatively more considerable way. Therefore, slime mould was mathematically formulated and utilized to solve optimization problems.

Mathematical Model Approach Food

The slime mould can find the food source using the odor in the air. The following equation is used to describe contraction mode:

$$X^{i+1} = \begin{cases} X_b^i + R1.(W.X_A^i - X_B^i) & if(r \geq p) \\ R2.X^i & Otherwise \end{cases} \tag{2}$$

$$p = \tanh|F_i - F^*| \tag{3}$$

where F_i represents fitness of X^i, F^* is the best fitness value obtained so far, X^i is the slime mould's position at $i-th$ iteration, $R1$ is a parameter generated in range $[-a, a]$, $R2$ is a parameter linearly decreases from 1 to 0, X_b^i and X_A^i are two randomly selected positions of slime mould, and W is the slime mould's weight.

$$a = \operatorname{arctan}h\left(-\frac{i}{iter_{\max}} + 1\right) \quad (4)$$

where $iter_{\max}$ is the maximum number of iterations.

$$W(SI^i) = \begin{cases} 1 + r.\log(\frac{F^* - F_i}{F^* - F_w} + 1) & condit \\ 1 - r.\log(\frac{F^* - F_i}{F^* - F_w} + 1) & Otherwise \end{cases} \quad (5)$$

$$SI = sort(F) \quad (6)$$

where r is a random number generated between zero and one, $condit$ denotes that F_i ranks initial half of the population, SI represents the fitness value sorted in ascending order, and F_w is the worst fitness value.

Wrap Food

The contraction behavior of venous tissue of slime mould is mathematically defined as follows:

$$X^* = \begin{cases} r.(U - L) + L & r < z \\ X_b^i + R1.(W.X_A^i - X_B^i) & if(r \geq p) \\ R2.X^i & r \geq p \end{cases} \quad (7)$$

where L is the lower boundary of the search space and U is the upper boundary of the search space.

Oscillation

Through the optimization process, $R1$ is randomly generated in $[-a, a]$ and reaches zero. The $R2$ is randomly generated from -1 to 1.

3 The Proposed Pistachios Species Classification Model

The overall proposed pistachios species classification model is comprised of three main phases as shown in Fig. 1; feature selection based on Slime Mould Algorithm (SMA) phase, features interpretation based on Explainable Artificial Intelligence (XAI) phases, and finally classification phase. In this first phase, SMA is used to select the optimal feature subset of twenty-eight features. Then the selected features obtained from the previous phase are interpreted and discussed using XAI in the second phase. Finally, in the classification phase, logistic regression is used to classify 2148 records of pistachios species. Logistic regression or shortly or logit regression is one of the statistical models that uses a logistic function to model a binary dependent

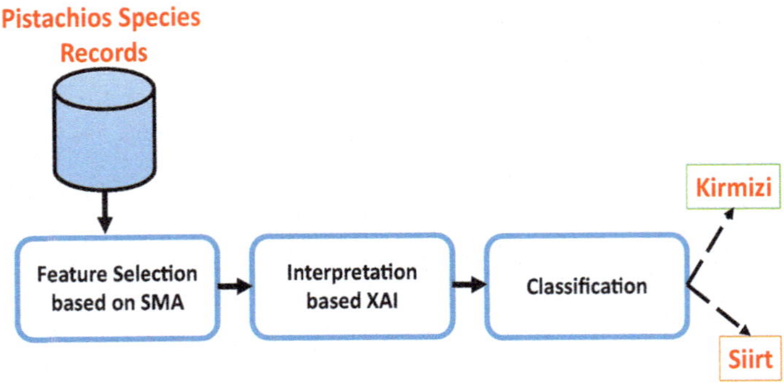

Fig. 1 The proposed pistachios species classification model

variable. In logit regression, the parameters of a logistic model are estimated as a form of binary regression. In this chapter, logit regression is used to classify pistachios species to either Kirmizi Pistachio or Siirt Pistachio. Next, a detailed description of the proposed SMA-based feature selection is proposed.

Slime mould algorithm starts with randomly initialization of each slim mould along with setting the other constant parameters; the maximum number of iteration ($iter_{\max}$) is set to 100, the population size (N) is set to 20, dimensionality size (dim) is set to 28, the search boundary ($[L, U]$) is set to [0, 1], and the control of the local and global search parameter (z) is set to 0.03. Each position of slim mould has different features with different sizes, where each element in the position vector contains either 0 or 1. The zero value in the vector means that the corresponding feature index isn't be selected and the one value means that this feature index will be considered. Thorough out the optimization process of SMA-based feature selection, each search agent is evaluated using a predefined fitness function. In this chapter, two main objectives are considered; the first one is to minimize the classification error and the second objective is to minimize the number of selected features, where KNN (K-Nearest Neighbour classifier) is used to evaluate how good is the slime mould position with K equals to 3 and Euclidean distance. To combine these two objectives in a single equation with use weights multiplied with each objective; the weight factor for the first objective is 0.99, while the weight factor for the second objective is set to 0.01. The mathematical formula of the used fitness function is defined as follows:

$$F_i = w1 * Error + w2 * \left(\frac{\sum_{j=1}^{dim} X_j^i}{dim} \right) \tag{8}$$

where F_i represents fitness of slime mould position X^i at i-th iteration, $w1$ equals to 0.99, and $w2$ equals to 0.01. The optimization process is repeated over and over till a termination criterion is satisfied. In this chapter, the algorithm terminates when it reaches $iter_{\max}$. The overall proposed SMA-based feature selection is introduced in the following algorithm.

Algorithm 1: The pseudocode of the proposed SMA based feature selection algorithm

Randomly initializes each positions of slime mould, set the constant parameters, $iter_{\max}$, N, dim, L, U, and z;
Evaluate each slime mould position using the fitness function as in Eq. (8);
Sort the faintness values and set F^* as the best fitness value obtained so far and F_w as the worst fitness value ;
while $i < iter_{max}$ **do**
    [Approach Food Phase] ;
    Update the weight of each slime mould using Eq. (5) ;
    Update the value of a parameter using Eq. (4);
    Update the value of p parameter using Eq. (3);
    Update the position of each slime mould using Eq. (2) ;
    Evaluate each slime mould position using the fitness function as in Eq. (8);
    if *F_i is better than F^** **then**
        Set F^* to F_i;
        Set X^* to X^i;
    end
    [Wrap Food Phase];
    Update the position of each slime mould using Eq. (7) ;
    Set i to $i + 1$;
end
return X^* and X^*;

4 Dataset Description

In this chapter, the dataset consists of 2148 records of pistachios species images [3]. The adopted dataset was categorized into two species, namely Kirmizi Pistachio and Siirt Pistachio. The dataset consists of twenty-eight features; twelve morphological features, four shape features, and twelve color features. The morphological features are Area, Perimeter, Major Axis, Minor Axis, Eccentricity, Eqdiasq, Solidity, Convex Area, Extent, Aspect Ratio, Roundness, and Compactness. While the shape features are Shapefactor-1, Shapefactor-2, Shapefactor-3, and Shapefactor-4. Finally, the color features are Mean-RR, Mean-RG, Mean-RB, StdDev-RR, StdDev-RG, StdDev-RB, Skew-RR, Skew-RG, Skew-RB, Kurtosis-RR, Kurtosis-RG, and Kurtosis-RB. Figure 2 shows a sample of the adopted dataset.

	A	B	C	D	E	F	G	H	I	J	K	L
1	Area	Perimeter	Major_Axis	Minor_Axis	Eccentricity	Eqdiasq	Solidity	Convex_Area	Extent	Aspect_Ratio	Roundness	Compactness
2	63391	1568.405	390.3396	236.7461	0.7951	284.0984	0.8665	73160	0.6394	1.6488	0.3238	0.7278
3	68358	1942.187	410.8594	234.7525	0.8207	295.0188	0.8765	77991	0.6772	1.7502	0.2277	0.7181
4	73589	1246.538	452.363	220.5547	0.8731	306.0987	0.9172	80234	0.7127	2.051	0.5951	0.6767
5	71106	1445.261	429.5291	216.0765	0.8643	300.8903	0.9589	74153	0.7028	1.9879	0.4278	0.7005
6	80087	1251.524	469.3783	220.9344	0.8823	319.3273	0.9657	82929	0.7459	2.1245	0.6425	0.6803
7	52268	1154.15	383.8558	197.502	0.8575	257.9723	0.8563	61039	0.5628	1.9436	0.4931	0.6721
8	71693	1463.932	388.2366	252.82	0.7589	302.1297	0.9156	78304	0.6892	1.5356	0.4204	0.7782
9	62240	1898.45	385.7238	218.1349	0.8247	281.5074	0.8947	69563	0.6727	1.7683	0.217	0.7298
10	64319	2011.291	435.9127	213.7594	0.8715	286.1704	0.8633	74502	0.6542	2.0393	0.1998	0.6565
11	78115	1238.623	492.33	204.9987	0.9092	315.3714	0.9616	81236	0.7105	2.4016	0.6398	0.6406
12	69391	1930.618	422.5333	223.0412	0.8493	297.2396	0.8948	77549	0.631	1.8944	0.2339	0.7035
13	67309	2353.405	451.3162	201.3479	0.895	292.7464	0.8906	75573	0.6887	2.2415	0.1527	0.6487
14	81656	1197.543	458.7472	237.3117	0.8558	322.4402	0.9164	89104	0.7482	1.9331	0.7155	0.7029
15	61725	1387.943	432.0616	194.6882	0.8927	280.3404	0.9016	68459	0.671	2.2192	0.4027	0.6488
16	81583	1295.3361	479.0619	228.7762	0.8786	322.296	0.928	87911	0.6971	2.094	0.611	0.6728
17	62862	1086.575	431.9981	188.4041	0.8999	282.9106	0.9662	65062	0.7527	2.2929	0.6691	0.6549
18	84537	1181.153	461.5339	235.7771	0.8597	328.079	0.9858	85753	0.7674	1.9575	0.7615	0.7108

Fig. 2 Sample of pistachios species dataset

5 Simulation Results and Discussions

The experimental results are divided into two main experiments. The first experiment aims to evaluate the performance of the proposed model using the Slime Mould Algorithm-based feature selection algorithm. The second experiment aims to evaluate the performance of the proposed pistachio species classification model using the Slime Mould Algorithm-based feature selection algorithm. In both experiments, Local Interpretable Model-Agnostic (LIME) is used to explain the prediction of logistic regression classifier. Several measurements are used for evaluation purposes. These measurements are accuracy, recall, sensitivity, precision, and F-score. It should be noted that all experiments are run on the same PC and same software. In this chapter, the used programming languages are Matlab R2020 and Python.

5.1 The Proposed Pistachio Species Classification Model without Using SMA

In this experiment, the performance of the proposed pistachio species dataset is evaluated. Figure 3 shows the explanation prediction output of the logistic regression classifier using LIME. In this figure, a random sample is used for interpretation. The

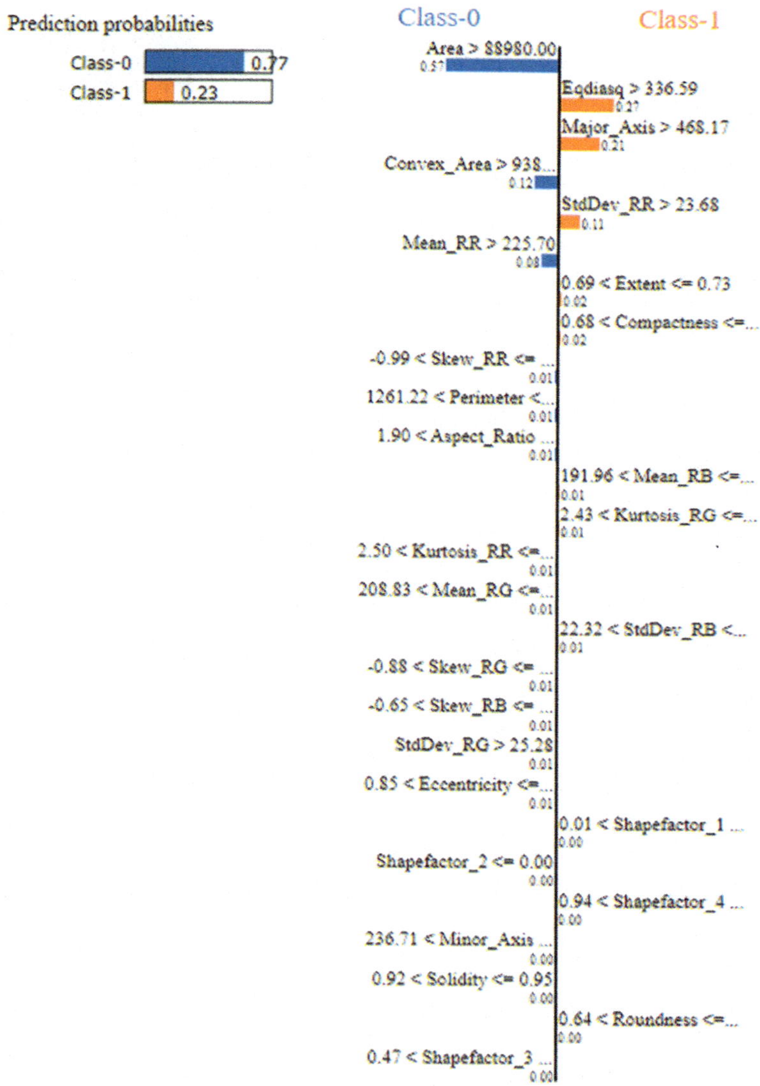

Fig. 3 Lime output for a random instance of pistachio dataset for whole attributes

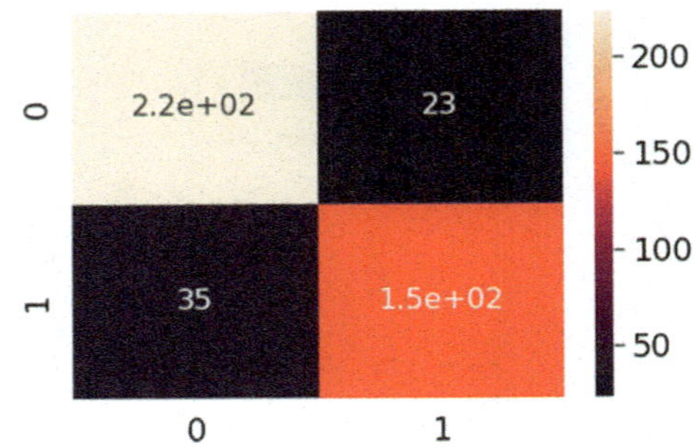

Fig. 4 Confusion matrix results of the proposed pistachio species classification model without using SMA

orange color indicates the contribution to the second class (class-1), while the blue color indicates the contribution to the first class (class-0). The first column showed the prediction probabilities of the two pistachios classes achieved from the logistic regression classifier. While the second column shows the contributions of features to the probability. As it can be observed, the logistic regression classifier predicts the instance with probability 77% for the first class and with probability 23% for the second class. Additionally, it can be seen that 11 features support the wrong class, while the other 17 features support the correct class. Thus, it can be concluded that most of these features are either redundant or noise and should be removed.

Figure 4 shows the confusion matrix of the proposed pistachio species classification model using all attributes. As it can be observed, there are many false-negative samples and false-positive samples. Thus, these results can affect the classification performance. The results from Fig. 4 are consistent with the explanation results obtained from LIME explainable artificial intelligence as shown in Fig. 3.

5.2 *The Proposed Pistachio Species Classification Model with Using SMA*

The second experiment aims to evaluate the performance of the proposed pistachio species classification model using the SMA-based feature selection algorithm. In this experiment, the selected features produced from the SMA-based feature selection algorithm are Aspect Ratio, Eqdisaq, Mean-RB, StDev-RRR, Mean-RG, and Minor Axis. Moreover, several evaluation criteria are used. These measurements, accuracy, precision, f-score, and confusion matrix.

Figure 5 shows the explanation prediction output of the logistic regression classifier using LIME. In this figure, the first column indicates the prediction probabilities, the second one indicates the feature probability, and the last one indicates the feature-value table. The feature probabilities graph indicates how many features influence the prediction results or the decision of a classifier machine learning algorithm. As it can

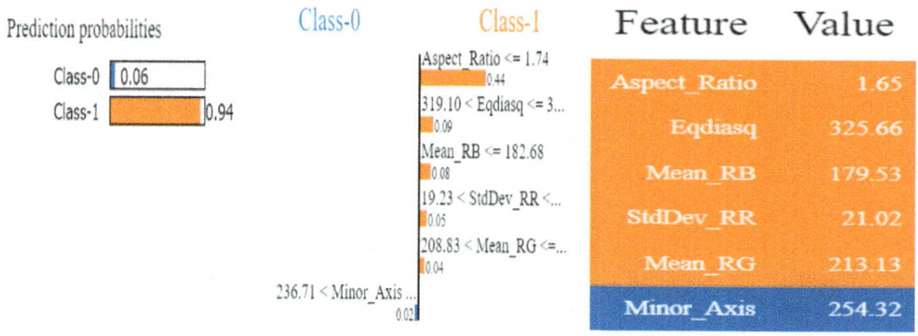

Fig. 5 Lime output for a random instance of pistachio dataset for selected attributes

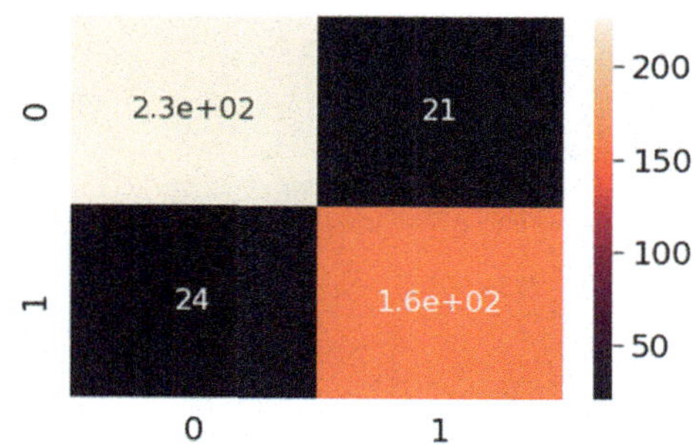

Fig. 6 Confusion matrix results of the proposed pistachio species classification model with using SMA

be observed, the logistic regression classifier predicts the instance with probability 6% for the first class and with probability 94% for the second class. Moreover, it can be seen that five features support the true class, while only one feature supports the wrong class. Thus, it can be observed that these selected features can be significantly used to distinguish between the two classes.

Figure 6 shows the confusion matrix of the proposed pistachio species classification model. As it can be seen, the selected feature is the dominant feature that can positively improve the logistic classification performance. Also, it can be observed that the true negative samples increases from 220 to 230, and the true positive samples from 150 to 160. While the false-negative samples decrease from 35 to 24, and the false-positive samples decreases as well from 23 to 21.

Figure 7 compares the performance of the proposed pistachio species classification model before and after applying the SMA-based feature selection algorithm. As it can be observed that working only with 6 features out of 28 can remarkably increase the accuracy percentage of the proposed pistachio species classification model. It obtained an overall accuracy of 90%.

Figure 8 compares the precision results of the proposed pistachio species classification model before and after applying the SMA-based feature selection algorithm. As the results show, the precision value of the proposed model after applying SMA is higher than the proposed model before applying SMA.

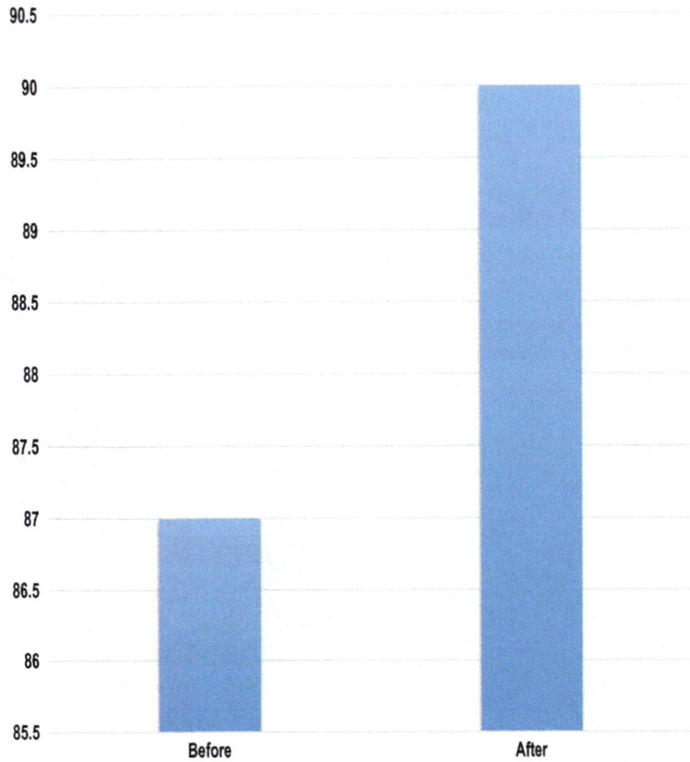

Fig. 7 Accuracy results before and after using SMA-based feature selection algorithm

Finally, Fig. 9 compares the performance of the proposed pistachio species classification model before and after applying the SMA-based feature selection algorithm in terms of f1-score. As it can be seen, the proposed model after applying SMA obtained an overall 91% f1-score. From all of the obtained results, it can be revealed that selecting only a subset of features and interpreting the results using LIME can significantly increase model interpretability and produce better results.

6 Conclusions

This chapter introduces the pistachio species, classification model. The proposed model consists of three main phases; SMA-based feature selection algorithm, interpretation based on XAI, and the classification and cross-validation phase. To get a

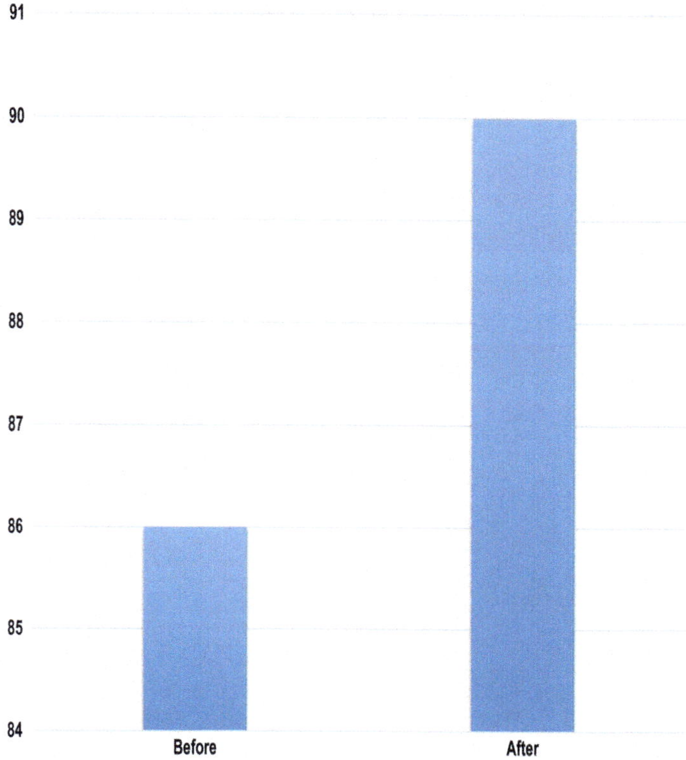

Fig. 8 Precision results before and after using SMA-based feature selection algorithm

better understanding and explanation of the overall pistachio species classification model, the Local Interpretable Model-Agnostic Explanations (LIME) xAI method is used. LIME is used to analyze a single sample of pistachio species dataset on a local level. The proposed model is applied to the dataset with 28 attributes and the number of records equals 2148. The experimental results demonstrated that using the LIME XAI method can assess the effectiveness of the proposed model and discover the main problem with a classifier. The proposed pistachio species classification model obtained overall 90% classification accuracy, 90% precision, and 91% f1-score.

In terms of future work, the proposed prediction model can be further applied to more complex and large datasets.

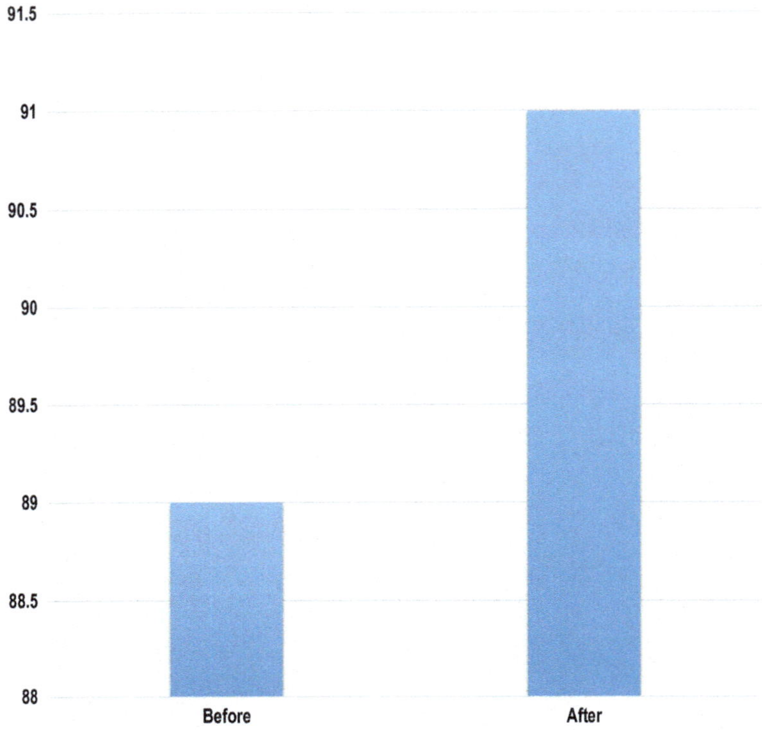

Fig. 9 F1-score results before and after using SMA-based feature selection algorithm

References

1. O. Bazrafshan, M. Ehteram, S. Dashti, Y. Feng, F. Yenn, A. Najah et al., Predicting crop yields using a new robust Bayesian averaging model based on multiple hybrid ANFIS and MLP models. Ain Shams Eng. J. **13**(5), 101724 (2022). https://doi.org/10.1016/j.asej.2022.101724
2. A. Heidary-Sharifabad, M. Zarchi, S. Emadi, G. Zarei, An efficient deep learning model for cultivar identification of a pistachio tree. Br. Food J. **123**(11), 3592–3609 (2021). http://doi.org/10.1108/BFJ-12-2020-1100
3. M. Koklu, Pistachio dataset (n,d.). https://www.muratkoklu.com/datasets/
4. S. Li, H. Chen, M. Wang, A. Heidari, S. Mirjalili, Slime mould algorithm: a new method for stochastic optimization. Future Gener. Comput. Syst. S0167739X19320941 (2020). http://doi.org/10.1016/j.future.2020.03.055
5. N. Mavani, J. Ali, S. Othman, Application of artificial intelligence in food industry-a guideline. Food Eng. Rev. **14**, 134–175 (2022)
6. S. Nosratabadi, S. Ardabili, Z. Lakner, C. Mako, A. Mosav, Prediction of food production using machine learning algorithms of multilayer perceptron and ANFIS. Agriculture **11**(5) 1–28 (2021). https://www.mdpi.com/2077-0472/11/5/408
7. I. Ozkan, M. Koklu, R. Saracoglu, Classification of pistachio species using improved K-NN classifier. Prog. Nutr. **23**(2) 1–9 (2021). http://doi.org/10.23751/pn.v23i2.9686

8. M. Ribeiro, S. Singh, C. Guestrin, "Why should I trust you?": explaining the predictions of any classifier, in *Proceedings of the 22nd ACM SIGKDD International Conference on Knowledge Discovery and Data Mining* (2016), pp. 97–101. https://doi.org/10.18653/v1/N16-3020
9. G. Sayed, A. Darwish, A. Hassanien, Binary whale optimization algorithm and binary moth flame optimization with clustering algorithms for clinical breast cancer diagnoses. J. Classif. **37**, 66–96 (2019). http://doi.org/10.1007/s00357-018-9297-3
10. G. Sayed, G. Khoriba, M. Haggag, A novel chaotic equilibrium optimizer algorithm with s-shaped and v-shaped transfer functions for feature selection. J. Ambient Intell. Hum. Comput. 1–26 (2022). https://doi.org/10.1007/s12652-021-03151-7
11. P. Tsakanikas, A. Karnavas, E. Panagou, A machine learning workflow for raw food spectroscopic classification in a future industry. Sci. Rep. **10**, 1–11 (2020)
12. Y. Zhang, Y. Weng, J. Lund, Applications of explainable artificial intelligence in diagnosis and surgery. Diagnostics **12**, 1–18 (2022). https://doi.org/10.3390/diagnostics12020237
13. L. Zhu, P. Spachos, E. Pensini, K. Plataniotis, Deep learning and machine vision for food processing: a survey. Curr. Res. Food Sci. **4** 233–249 (2021). http://doi.org/10.1016/j.crfs.2021.03.009

A Support Vector Machine Model for Rice (*Oryza sativa* L.) Leaf Diseases Based on Particle Swarm Optimization

Walid Hamdy, Amr Ismail, Wael A. Awad, Ali H. Ibrahim, and Aboul Ella Hassanien

Abstract Our society is becoming increasingly reliant on technology every day. Agriculture, on the other hand, is critical to human survival. Rice is one of the most important grains of food. It feeds nearly half of the world's population and supports a large number of jobs. Crop diseases cause significant yield losses around the world, particularly in Sub Saharan Africa. To implement effective disease management measures, pathogen detection and understanding spatiotemporal dynamics are critical, and this necessitates the use of molecular detection methods, particularly to discriminate between infections that cause similar symptoms. As a result, adequate disease mitigation for rice plants is critical. We used to our proposes a model for detecting three rice leaf diseases: brown spot, leaf smut, and bacterial leaf blight. This paper presents a unique model for the classification of rice leaf diseases by multi-class SVM, K—means clustering, and PSO. The disease was classified using an SVM classifier and the classifier accuracy is optimized using PSO. The exploratory results show that the proposed approach performed well in terms of disease detection accuracy, with a score of 98.89%.

W. Hamdy · A. Ismail · A. H. Ibrahim
Faculty of Science, Port Said University, Port Said, Egypt
e-mail: walid.hamdy@ckes.cu.edu.eg
URL: http://www.egyptscience.net

W. Hamdy
Modern Academy for Computer Science and Management Technology, Cairo, Egypt

W. A. Awad
Faculty of Computers and Artificial Intelligence, Dameitta University, Dameitta, Egypt
e-mail: Wael_abdelkader@du.edu.eg

A. E. Hassanien (✉)
Faculty of Computers and Artificial Intelligence, Cairo University, Giza, Egypt
e-mail: aboitcairo@cu.edu.eg
URL: http://www.egyptscience.net

W. Hamdy · A. E. Hassanien
Scientific Research Group in Egypt (SRGE), Cairo University, Cairo, Egypt

A. E. Hassanien and M. Soliman (eds.), *Artificial Intelligence: A Real Opportunity in the Food Industry*, Studies in Computational Intelligence 1000,
https://doi.org/10.1007/978-3-031-13702-0_4

Keywords Image classification · Convolutional neural network · Plant leaf diseases · Rice leaf diseases

1 Introduction

Rice (*Oryza sativa L.*) is one of the most important crops in the world, as it is the primary food for more than three billion people. Bangladesh, China, India, Indonesia, Vietnam, and Nigeria all eat it as a staple cuisine [1]. Anything that has an impact on rice crop quality and quantity has an impact on the entire population. As a result, illness identification and monitoring must be done on a regular basis to ensure effective mitigation. Diseases can wreak havoc on production if they aren't treated quickly. As a result, in agro-informatics, automatic plant disease identification is a hot topic. Some of the most common rice plant diseases include bacterial leaf blight (BLB), brown spot (BS) and leaf smut (LS) [2]. Affected plants produce less rice and rice of lesser quality than healthy plants. This decrease in yield has a direct and detrimental impact on the general population, food security, and the national economy. Bacterial leaf blight can lower yield by up to 70%, according to the International Rice Research Institute [3]. Rice production can be reduced by up to 45% due to brown spot disease [4]. It was one of the key causes of the 1943 "Great Bengal Famine." Some of these illnesses have identical symptoms, which can cause confusion among inexperienced farmers and result in harvest loss [5]. This productivity loss may have been averted if there had been a simple, automated, and reliable illness detection facility. Such a strategy will almost certainly provide tangible effects in a short period of time [6]. We aim to develop a set of useful features in this research that will help us classify rice leaf diseases more accurately.

2 Related Works

Nowadays, the popular study topic is the implementation of information technology and computers in the agricultural domain. Different researchers are involved in various aspects of this topic. Garcia et al. [7] reviewed several classification strategies and image processing in this sector, while Prajapati et al. [8] reviewed numerous rice plant disease detection strategies. Shimamura et al. [9] used CNNs to automate tipburn detection in artificially lighted plant factories. Sugimoto et al. [10] looked on the yield loss caused by weeds in rice fields. Robotics were used to design a paddy weed removal technology. Many academics have worked on the topic of automatically detecting plant illnesses from leaf photos. In their method, Khirade et al. [11] for image segmentation, they employed hue-saturation thresholding, and for classification, they used SVM. Wang et al. [12] used a dataset with two wheat leaf classes and two grape leaf classes to extract 50 characteristics. For disease classification, they employed K-means using a backpropagation neural network for segmentation and PCA dimensionality reduction. Back propagation neural networks using statistical features from a color picture dataset encompassing Blast, bacterial leaf blight, and Brown Spot illnesses were also used by Orillo et al. [13]. Suman et al. [14] studied diseases such as bacterial leaf blight, narrow brown spot, brown spot, and rice blast. They employed SVM for classification and 8-connected component analysis for illness segmentation. They were able to extract a variety of color and shape attributes for their model, which was 70% accurate. Singh et al. [15] employed K-means clustering for segmentation of impacted spots. They employed SVM to classify the data and achieved an accuracy of 82%. Sethy et al. [16] also utilized K-means in their analysis. Using SVM as the classifier, shape and texture features were incorporated by Yao et al. [17]. Rice blast and brown spot infections were studied by Phadikar et al. [18] Color and form features such as hue radial distribution were extracted. For segmentation, as a classifier, they used a membership function and a mathematical morphology. They achieved over 70% accuracy using color, shape, and texture parameters. In the work of Chawathe [19], several color and form features were retrieved as features [20].

3 Materials and Methods

This section delves into the specifics of the CNNs used to diagnose plant diseases on the Rice leaves dataset in this study. The goal of this study is to optimization CNN model that has already been trained using PSO optimization. Data acquisition, data training, data categorization, and data assessment are the four essential phases of the process, which are explained below.

3.1 Dataset Description

Prajapati et al. [21] created the dataset used in this study, which contains 120 RGB images divided into three classes: brown spot, leaf smut, and bacterial leaf blight, each with 40 photos. Images are taken with natural lighting on a white background. BLB illness is caused by the bacteria *Xanthomonas oryzae*, which is found in rice. A hot, humid environment is ideal for the transmission of disease. That is why it is so common in rice-growing regions of Asia, Africa, and the Caribbean. Crop loss might be as high as 75% during a BLB epidemic. It can harm rice leaves and seedlings, seeping milky fluid at first and then leaves drying, causing them to droop and die. The cause of brown spot disease by the fungus *Cochliobolus miyabeanus*. It affects at leaves of the rice plant, sheath, stems, and seedlings, among other things. The yield loss can be as high as 45% or higher. The diseased patches on paddy leaves are oval or circular in shape. It causes the plant's damaged section to die, reducing the quality and quantity of the yield. *Entyloma oryzae*, a fungus, is also responsible for leaf smut (LS). Although it is not a severe disease, it can help the spread of other diseases by providing a suitable environment for the fungus. Lesions range in shape from round to oval or irregular and have a rough texture.

Figure 1 illustrates a few representative photos from the dataset of various disorders. Extracting significant characteristics is necessary for developing a reliable classification model. Table 1 summarizes the most important aspects of the disorders for this purpose. These color and shape data can be processed to develop unique characteristics.

Fig. 1 Rice leaves with various diseases; **a** bacterial leaf blight, **b** leaf smut, **c** brown spot

Table 1 Summarizes the most important aspects of the disorders for this purpose

Name of disease	Symptom color	Lesion shape
– Bacterial leaf blight – Leaf smut – Brown spot	– Yellow to Gray–White – Dark Reddish Brown to Dark Gray – Light Brown to Gray at the center, Reddish Brown surrounding	– Long spots, non-uniform shape – Circular to oval or non-uniform – Circular to oval

3.2 Proposed Method

In this paper, a mechanism for classifying Rice diseases such as bacterial leaf blight (BLB), brown spot (BS), and leaf smut is proposed (LS). The methodology shown in Algorithm 1 reveals the presented system. Figure 2 depicts the planned work's flowchart.

Algorithm 1 The methodology for our model
1: Acquire images from data set
2: Pre-process the images for noise removal, resizing, and contrast enhancement
3: Separate images among train and test data
4: Perform k-means algorithm with PSO algorithm
5: Feature extraction
6: Training the SVM classifier for optimized value = 0

The proposed approach contained three main stages: preprocessing algorithms, The PSO algorithm, and finally classification and evaluation. The suggested proposed system's overall architecture is detailed below:

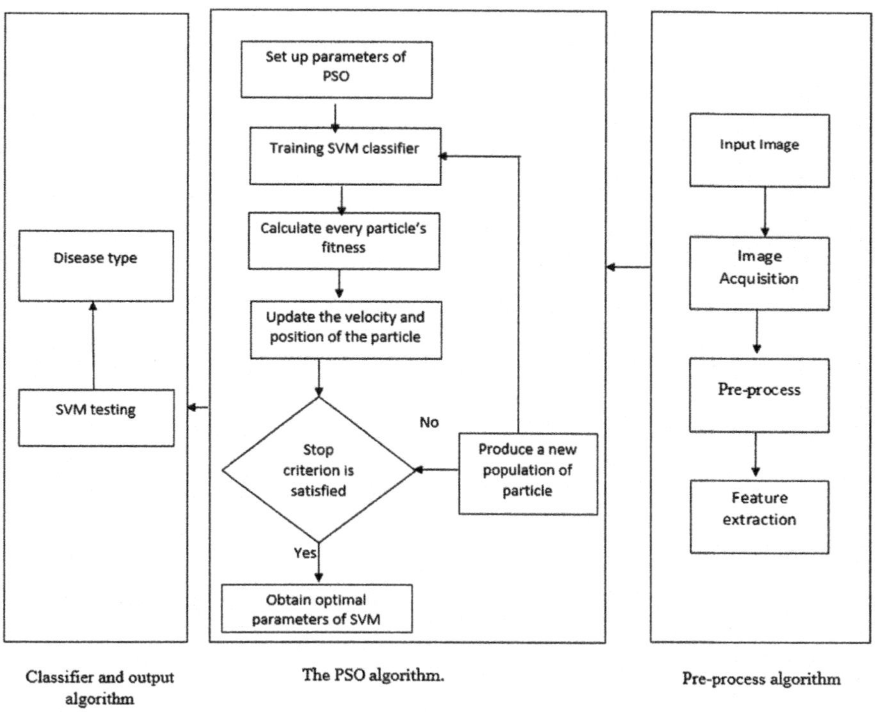

Fig. 2 Flowchart of the proposed work

Firstly the Processing Algorithms

The input image is in RGB form. The images are pre-processed for noise removal, resizing, and lightning improvement in our proposed model after the dataset is inserted.

Resizing All of the images of our dataset were resized by using the nonadaptive model of interpolation. K-nearest neighbour used for resized the input images in our approach.

Noise Removal Depending on the type of noise, different filters are used to remove it. Some of the filters used to reduce impulsive noise in color images are Gaussian filters, Vector median filters, and hybrid filters. The median filter is used in the proposed work because it has the property of preserving edges even after the noise has been removed.

Histogram Equalization Equalization of the histogram spreads out the most common intensity levels. To improve visual contrast, a uniform intensity value is assigned to each pixel used the histogram of a picture and the histogram equalization approach provided by Eq. 1.

$$G\left(P_{(a,b)}\right) = round\left(\frac{f_{cde}\left(p_{a,b}\right) - f_{cdemin}}{(1 * A) - f_{cde}} * H - 1\right) \tag{1}$$

Second: Particle Swarm Optimization Algorithm (PSO)

PSO is a population-based stochastic optimization method inspired by the social behavior of birds flocking or fish schooling, and was created by Russell Eberhart and James Kennedy in 1995. Genetic algorithms or evolutionary computation are comparable to PSO. Particles are a group of individuals that travel in phases throughout a location. The objective function for each particle is assessed at each phase. The algorithm determines the new velocity of the particle based on this evaluation. This process is repeated until an optimal solution is found. PSO is a simple, easy to implementing programming language that has been employed in a variety of applications. Equations 2 and 3 describe the particle's movement:

$$V_s(t+1) = \omega V_s(t) + c1r1(P_s - X_s(t)) + c2r2(G - X_s(t)) \tag{2}$$

$$X_s(t+1) = X_s(t) + V_s(t+1) \tag{3}$$

where Vs(t) = the s particle velocity in t time, Xs(t) = position of particle s vector in time t, Ps = personal best particle s solution, G =the best particle solution has been discovered at present, ω = inertia weight, $c_1 = c_2$ = acceleration constant known as cognitive and social parameters, $r_1 = r_2$ =random functions in the range [0, 1].

Finally Classification and Evaluation

In this phase, the leaf diseases of rice crops were classified and compared by recording the relevant feature set of values in their respective datasets. Support vector machines (SVM) with high dimension spaces, fast memory, and versatile decision function are employed here as a completely controlled learning method for categorization. The extracted feature dataset is first optimized with PSO, and then the classification process is carried out with multiclass SVM. The amount of cataloguing is done through high scale percentage of classification gain, Equation 4.

$$Classification\,Gain(\%) = \frac{Number\ of\ correct\ classification}{Total\ no\ of\ test\ images} * 100 \qquad (4)$$

For our classification using SVM+PSO the data split into two sets of the proposed model is done in two method:

- Train data the machine with the train dataset (using 75% of the images data).
- Test date the model (using 25% of the images data).

3.3 *Results*

As input, various types of sick rice leaf samples are used. This research examines three types of rice leaf disease: bacterial leaf blight, brown spot, and leaf smut. The entire amount of photos trained in the data set paddy at distinct classes is referred to as the image set. The total number of diseased photographs (3 classes) is 120. A total of 90 images were used as training samples. (There are 30 photos in each class) The total number of testing samples is 30. (each class contains 10 images). In every scenario studied, the results achieved utilizing the established PSO + SVM technique were superior to those obtained using grid search. Without compromising SVM classification accuracy, the PSO + SVM technique may simultaneously determine parameter values and discover a subset of features. Table 2 compares the results of our methods to those of other model.

We can be seeing that SVM + PSO enhances performance over SVM models, demonstrating the SVM+ PSO model's usefulness. In comparison to other models, SVM+PSO achieves the highest results in five tasks and is very competitive in the

Table 2 The results obtained from the compare models between SVM model and SVM + PSO model

Epoch number	SVM	SVM + PSO
Epoch 1	94.59	98.35
Epoch 2	95.02	98.89
Epoch 3	93.98	97.69
Epoch 4	92.59	97.97
Epoch 5	93.21	98.35

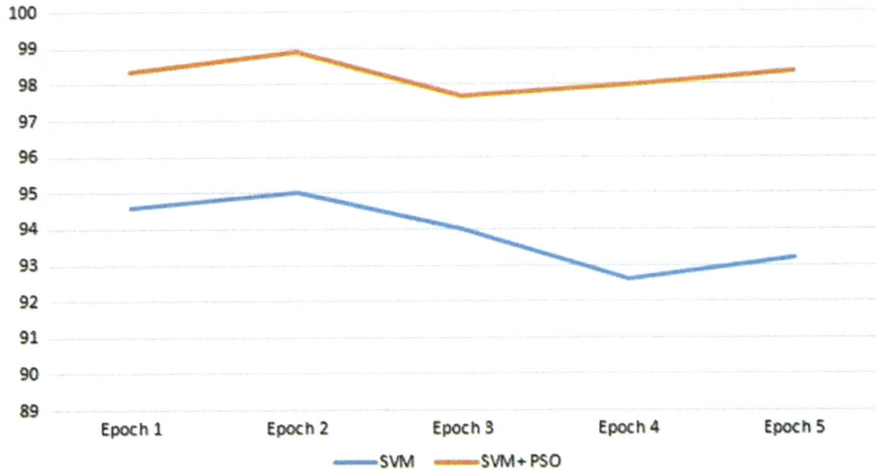

Fig. 3 Accuracy curves for compare between the SVM model and SVM + PSO model

others. We show the curves of the better accuracy model found by SVM+ PSO on the convex dataset for 12 epochs in comparison to the SVM model in Fig. 3. We can observe that our model's accuracy has improved over the SVM model, implying that the suggested SVM+ PSO is genuinely capable of discovering good models for a particular image classification dataset.

Accuracy is defined as the proportion of correct forecasts to all forecasts made, expressed as a percentage and quantified using Eq. 5.

$$Accuracy = \frac{Correct\ prediction\ number}{Total\ prediction\ number} \tag{5}$$

4 Concluding Remarks

In this study, we present a method for classifying rice leaf diseases. Leaf Smut, Leaf Blight, and Brown Spot are three rice leaf diseases that can be detected with this method. When it comes to classifying the rice leaf disease, the SVM classifier has a 95.02% accuracy rate. On the basis of a comparison between the suggested PSO with SVM algorithm and the SVM algorithm, we evaluated the performance of the suggested PSO-SVM approach. It has also been demonstrated that the proposed PSO-SVM approach can be used to verify the productivity of planned approaches. SVM demonstrated to be a promising strategy for the differentiation and categorization of rice leaf diseases with a normal precision of 98.89% after utilizing the

PSO methodology to optimize the feature dataset. The study could be expanded by collecting a larger number of samples from a wider range of damaged rice leaves.

References

1. International Rice Research Institute, "Bacterial blight-IRRI Rice Knowledge Bank," Rice Knowledge Bank. [Online]. Available: http://www.knowledgebank.irri.org/training/fact-sheets/pest-management/diseases/item/bacterial-blight. Accessed 28 Nov 2019
2. N. Nandhini, R. Bhavani, Feature extraction for diseased leaf image classification using machine learning. International Conference on Computer Communication and Informatics (ICCCI) (2020), pp. 11–4
3. A. Elaraby, W. Hamdy, M. Alruwaili, Optimization of deep learning model for plant disease detection using particle swarm optimizer. CMC-Comput. Mater. Continua **71**(2), 4019–4031 (2022)
4. International Rice Research Institute, "Brown spot-IRRI Rice Knowledge Bank," Rice Knowledge Bank. [Online]. Available: http://www.knowledgebank.irri.org/training/fact-sheets/pest-management/diseases/item/brown-spot. Accessed 28 Nov 2019
5. W. Hamdy, I. Elansary, A. Darwish, A.E. Hassanien, An optimized classification model for COVID-19 pandemic based on convolutional neural networks and particle swarm optimization algorithm. Digital Transformation and Emerging Technologies for Fighting COVID-19 Pandemic: Innovative Approaches, (2021) pp. 43–61
6. W. Hamdy, I. Elansary, A. Darwish, A.E. Hassanien, Artificial intelligence strategy in the age of Covid-19: opportunities and challenges. Digital Transformation and Emerging Technologies for Fighting COVID-19 Pandemic: Innovative Approaches (2021), pp. 81–93
7. J.G.A. Barbedo, Digital image processing techniques for detecting, quantifying and classifying plant diseases. Springerplus **2**(1), 1–12 (2013)
8. J.P. Shah, H.B. Prajapati, V.K. Dabhi, A survey on detection and classification of rice plant diseases. In 2016 IEEE International Conference on Current Trends in Advanced Computing (ICCTAC). IEEE (2016), pp. 1–8
9. S. Shimamura, K. Uehara, S. Koakutsu, Automatic identification of plant physiological disorders in plant factories with artificial light using convolutional neural networks. Int. J. New Comput. Architect. Appl. **9**(1), 25–31 (2019)
10. M. Sugimoto, K. Takeuchi, T. Yamaji, H. Inoue, K. Hosotani, Y. Inoki, D. Yoshioka, M. Endo, M. Kato, T. Shirakawa, H. Fukita, A study for development of autonomous paddy-weeding robot system-an experimentation for autonomously straight-running based on compass-compensation. Int. J. New Comput. Architect. Appl. **8**(4), 186–198 (2018)
11. S.D. Khirade, A.B. Patil, Plant disease detection using image processing. In 2015 International Conference on Computing Communication Control and Automation. (IEEE, 2015), pp. 768–771
12. H. Wang, G. Li, Z. Ma, X. Li, Image recognition of plant diseases based on backpropagation networks. In 2012 5th International Congress on Image and Signal Processing . (IEEE, 2012), pp. 894–900
13. J.O. William, J.D. Cruz, L. Agapito, J.S. Paul, I. Valenzuela, Identification of diseases in rice plant (*Oryza sativa. L*) using back propagation artificial neural network. In International Conference on HNICEM (2014), pp. 1–6
14. T. Suman, T. Dhruvakumar, Classification of paddy leaf diseases using shape and color features. IJEEE **7**(01), 239–250 (2015)
15. A.K. Singh, A. Rubiya, B.S. Raja, Classification of rice disease using digital image processing and svm classifier. Int. J. Electric. Electron. Eng. **7**(1), 294–299 (2015)
16. P.K. Sethy, B. Negi, N. Bhoi, Detection of healthy and defected diseased leaf of rice crop using K-means clustering technique. Int. J. Comput. Appl. **157**(1), 24–27 (2017)

17. Q. Yao, Z. Guan, Y. Zhou, J. Tang, Y. Hu, B. Yang, Application of support vector machine for detecting rice diseases using shape and color texture features. In 2009 International Conference on Engineering Computation (IEEE, 2009), pp. 79–83
18. S. Phadikar, J. Sil, A.K. Das, Classification of rice leaf diseases based on morphological changes. Int. J. Inf. (2012)
19. S.S. Chawathe, Rice disease detection by image analysis. In 2020 10th Annual Computing and Communication Workshop and Conference (CCWC) (IEEE, 2020), pp. 0524–0530
20. A. Elaraby, W. Hamdy, H. Nisar, M.H. Alkinani, Segmentation of Activated Sludge Flocs in Microscopic Images for Monitoring Wastewater Treatment. Complexity (2022)
21. H.B. Prajapati, J.P. Shah, V.K. Dabhi, Detection and classification of rice plant diseases. Intell. Decision Technol. **11**(3), 357–373 (2017)

CNN in Food Industry: Current Practices and Future Trends

B. K. Tripathy, A. Salaria, and A. Ambastha

Abstract The food industry is a multifaceted, global net of different commerce that conveyances most of the food disbursed by the world's inhabitants. The word 'Food Industries' shelters a chain of trade deeds concentrating on the manufacture, circulation, dispensation, alteration, preparation, preservation, transportation, endorsement and packaging of food products. Technologies such as Artificial intelligence that deal with constructing clever machinery adept at performing jobs that characteristically necessitate human astuteness are now being used in all different industries together with the food industry all over the world. With food being a never ending and ever increasing necessity of every nation, AI has a prodigious opportunity in the food industry in the upcoming decades. In this chapter, we shall be deliberating some applications of AI, primarily focused on Convolutional Neural Network (CNN) that are at present being used in the food industry, challenges that are being confronted right now in the present-day technology and most notably, some of the future applications of CNN in the food industry.

Keywords Artificial intelligence (AI) · Convolutional neural networks (CNN) · Food industry · Applications of CNN in food industry · Future trends

1 Introduction

Food is a basic necessity for survival. With human evolution, the food industry started growing and continues to evolve. It poses as a vital industry that will never be obsolete. When information technology is fused with the food industry, it reaps many benefits. Not only will it facilitate lesser manual help, but will also provide for an increase in scope of newer opportunities in the industry. Currently, India's food sector contributes to about 14% of manufacturing GDP, nearly 13% of India's exports, and about 6% of the industry's total investment. Food processing is a big business in India and has the potential to be bigger in the coming years. Hence, a

B. K. Tripathy (✉) · A. Salaria · A. Ambastha
School of Information Technology and Engineering, VIT, Vellore, India
e-mail: tripathybk@vit.ac.in

A. E. Hassanien and M. Soliman (eds.), *Artificial Intelligence: A Real Opportunity in the Food Industry*, Studies in Computational Intelligence 1000,
https://doi.org/10.1007/978-3-031-13702-0_5

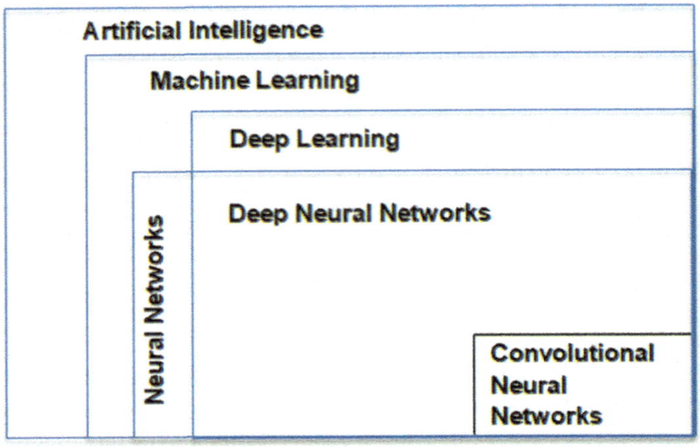

Fig. 1 Artificial intelligence and its sub categories

need for a smarter solution that aids and suggests improvements in the field is the need of the hour.

Artificial Intelligence (AI) is a broad category of methods which has machine learning (ML) as a part of it. Basic aim of AI is to mimic human brain by the design of systems which are capable of exhibiting characteristics of human brain shown through their intelligence. It has wide spread applications in various aspects of human society and handles processes like understanding language, reasoning, learning and taking actions. John McCarthy, who coined the term "Artificial Intelligence", introduced it in a conference held in the year 1956 and introduced it as "the science and engineering of creating intelligent machines". AI technology has now become a vital part of the digital revolution.

Due to the amplified demand and enhancing mandate in the food industry that might be the reason industry has opened its door to cuddle AI technologies to explore new ways to reach and serve consumers in order to extract best profits out of it. AI has been efficaciously deployed for time saving, best utilization of resources with greater efficiency through different applications starting from production until it reaches the consumers. Although, AI has some challenges but its applications benefits overpower its challenges in food industry [6].

AI and its subfields are depicted in the following diagram (Fig. 1).

1.1 Machine Learning

ML is a branch of AI where machines learn from data provided without much human intervention. The machines can discover to perform certain functions without being programmed to do so. In this branch of science systems of computers and machines try to process the input data, make their analysis and finally interpret so that solutions for real life problems can be generated. Through ML techniques and algorithms a model can be trained and tested for accuracy so that futuristic events can be predicted. Basically the system developed gets trained and tested for future use. Under these processes the system learns by itself without much human intervention. This can be considered as learning by example. The previous occurrences help the system to generate some approach, may be in the form of rules and predict outcomes. It is different from the program driven feature of the systems which cannot do anything beyond the directions in the programs. Some examples of this technology are automatic vehicles, speech recognition and web search, speech, which are only a few applications of Machine Learning.

Broadly the learning processes under the umbrella of Machine Learning can be categorized as:

- Unsupervised Learning
- Supervised Learning
- Reinforcement Learning.

1.2 Artificial Neural Networks

An Artificial Neural Network (ANN), which is either an algorithm or an actual hardware, is a processing device designed after the design of animal brains and their functioning has the ability to learn by example. These are based upon and are inspired by central nervous system of humans. ANNs have enough flexibility, suitable for real-time systems; have fast response(less computational time) and support parallel architecture. It is well known that human brain is a large mesh of biological neurons which are interconnected. Similarly, ANNs comprise of simple signal processing elements called artificial neurons which are connected following specific patterns. Many basic functions of the brain like pattern recognition and data classification of the brain are emulated by the ANN [39].

The neurons are arranged vertically in layers. Neurons of a layer are connected to each of the neurons in the succeeding layers in a fully connected ANN. The connections are directed arrows and attached with weights. The numbers of layers, number of neurons in a layer along with the connections form the architecture of an ANN. There are two specific layers called the input layer (where the signals are received) and the output layer (where the final output signals come out) form the mandatory components and together with the interconnections are called a single layered ANN. The functions of each neuron in an ANN are specified.

In Fig. 2, there are two neurons X1 and X2 in the input layer and a single neuron in the output layer. The input signals are represented by $\times$ 1 and $\times$ 2, whereas the output signal is represented by 'y'. The weights of the interconnections X1 and X2 to Y are represented by w1 and w2 respectively.

The total input is represented by y_{in} and is given by

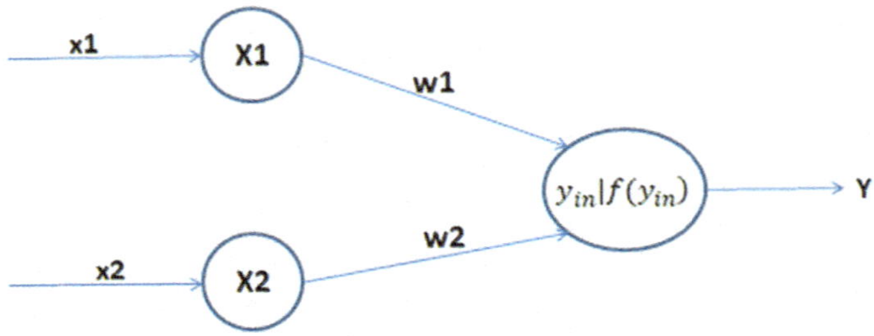

Fig. 2 A simple ANN model

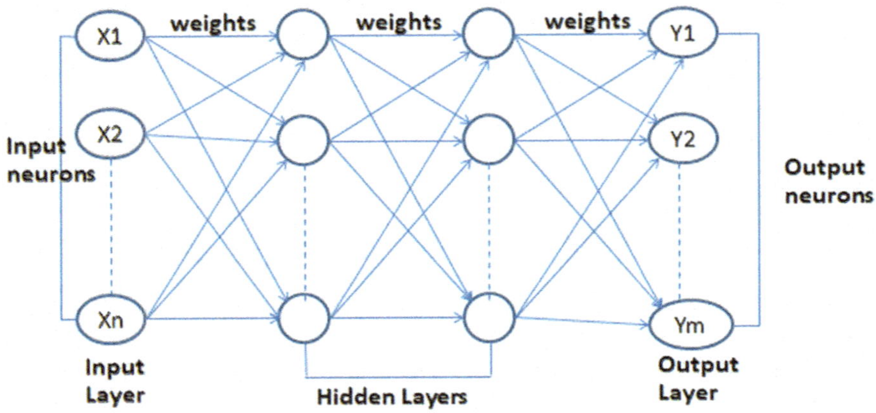

Fig. 3 Architecture of a multi layered Feed forward Neural Network

$$y_{in} = w_1x_1 + w_2x_2 \tag{1}$$

A function called the activation function is applied to y_{in} to generate the output signal 'y'.

There may be more than one neuron in the output layer and also there may be a few intermediate layers in between the input and output layers. These intermediate layers are called the hidden layers. Each hidden layer increases the complexity of the ANN but makes the computation more accurate also. A multilayered feed forward ANN is presented in Fig. 3.

1.3 Deep Learning and Deep Neural Networks (DNNs)

It is a process of hierarchical feature learning in which the features of lower layers in the higher layers are formed by composing the features in the lower layers. The features are automatically learnt and it happens to be a strong characteristic to learn complex functions. During this outputs are learnt form data with very little features being designed by humans. It is important to note that high level abstractions are obtained as a result and even humans are not aware of the high level abstractions specifically.

The process of human decision making is followed in Deep learning. It is a category of AI [4] Neural networks are the backbone of deep learning algorithms. In Fig. 1 a diagrammatic representation of the branches of knowledge AI, ML, Deep learning, neural Networks and Deep neural Networks is illustrated. The Deep learning architecture forms a hierarchy of features such that the features at a higher level are a composition of the features at the next lower level. The number of levels of nonlinear operations which form the parts of the function to be learned defined the depth of the architecture. The transfer of data is done from the input values to the output values through transformations which depict their relationships. Deep Learning is related with the transformation and extraction of features which endeavours to build a relationship amid stimuli and connected neural responses existing in the human brain.

Deep learning [24] refers to specific types of machine learning algorithms where numerous layers of uncomplicated processing units are linked in a network in such a way that the input to the system is approved through every one of them one after the other. This architecture has been enthused by the processing of visual info in the brain approaching through the eyes and apprehended by the retina. The depth permits the network to learn further complex structures without necessitating unrealistically enormous amounts of data [2]. The ability of deep learning to work easily with large datasets is what makes it trending in today's big data era. The capability to process big amounts of characteristics makes deep learning very influential when coping with unstructured data [5].

The human brain architecture is deep hierarchical in structure. Inspired by the processing of concepts in human brain, which is hierarchical by nature with several stages of transformation and representation, neural networks scientists have tried for several years for training deep multi-layered neural networks. This search was not successful till the development of Deep Belief Networks (DBNs) by Hinton et al. at the University of Toronto in 2006 [25]. The training with 2 or 3 hidden layers were resulting in positive outcomes. But, beyond that the results obtained were of lesser intensity. The DBFs use the method of training one layer at a time (greedy process), which is the essence of a unsupervised learning procedure called the Restricted Boltzmann Machine (RBM) [17]. Shortly after that autoencoders were developed, which are typically based upon the principle where the intermediate layers are trained locally with their levels being represented by unsupervised learning. This has led to the development of several Deep Neural Network (DNN) models [9], all of which use the same principle mentioned above and are applied to many tasks such as classification, dimensionality reduction, regression, object segmentation, modeling textures and information retrieval. Most of these applications have successfully used deep supervised feedforward neural networks intended for the individual tasks.

A class of ANNs derived from feed forward networks is the recurrent neural network (RNN) in which the nodes are arranged in the form of directed circuits and a variable length of sequences of inputs are processed using internal state.

A form of RNN used in the area of deep learning is the Long short-term memory (LSTM), where in addition to feedforward connections there are feedback connections also. It can process sequences of data like those in video or speech, connected handwriting recognition [3]. One of the latest variants of DNN is the GAN [1]. In the framework of a GAN, two models called the generative model and discriminatory model are used. The discriminatory model checks if a sample is from the generative model or not and the generative model tries to produce fake samples which cannot be identified by the discriminatory unit. During this competition both the units try to improve until the fakes cannot be distinguished from the genuine ones.

1.4 Convolutional Neural Networks (CNNs)

Convolutional Neural Networks (CNNs) [29] are constructed diversely as compared to an original neural network. The standard neural networks have a fully connected architecture. Whereas in a CNN the fully connectedness need not be there. It has width, height and depth as the 3-dimensional layers. In such an architecture each node in a layer is connected to a group of neurons in the layer below it, which is normally a small subset of the nodes in the layer below [26].

The problem of vanishing gradients has been a troublesome problem in the realm of backpropagation networks. In an attempt to resolve this and fins a fruitful solution the model of Residual Network (ResNet) was proposed by researchers at Microsoft Research in 2015 [13]

This chapter will discuss CNN, an extraordinary kind of ANN, precisely its applications in the food industry. In the next section, we give a brief introduction to CNN. In Sect. 3, we will be discussing some of the major applications of CNN popular today and also about the current practices of CNN in the food industry and finally, we will give some of the future trends related to CNN in the food industry.

2 CNN Architecture

It is difficult for businesses to incorporate convolutional neural networks and neural networks into real time applications. There are manifold factors such as what convolutional architecture is to be used, what sort of datasets are to be used, what deep learning model is to be used to accommodate the data etc. that need to be taken into consideration to integrate CNN with deep learning.

It is preferable to map the chief architectures of networks offered by deep learning with key architectures of Convolutional Neural Networks (CNNs). Businesses could conceivably embrace the approach of transfer learning to form a set of images and then train the designated network architecture in the determined dataset. Fundamentally, for a trouble-free amalgamation (i) choosing a CNN architecture which is capable of modeling data alike to the data that one wants to model to determine that it is proficient of modeling the anticipated features, (ii) a system that permits formation and distribution the model within the restrictions of data storage and security and (iii) managing the admittance to the precise version of each model across numerous dissimilar applications are a must.

2.1 CNN Layers

CNN is constructed with three kinds of layers, namely, the convolutional layers, pooling layers, and fully-connected (FC) layers [20]. When these layers are piled up, CNN architecture is established (Fig. 6).

The Convolutional or Math Layer is the first layer that is used to excerpt the numerous characteristics from the input images. It is called the Math Layer because all the mathematical calculations of intricacy are implemented amid the image presented and a filter of a specific size. The scalar product is calculated amid the filter and the parts of the image presented with regard to the filter size by slipping the filter over the image presented.

The Pooling Layer follows the Math Layer. The Pooling Layer generally functions as a conduit between the Convolutional Layer and the Fully Connected (FC) Layer. The chief goal of this layer is the diminution of the size of the convolved feature map to lessen the computational expenses. This is achieved by reducing the associations between layers and autonomously operating on each feature map (Fig. 4). Reliant upon technique used, there are many kinds of pooling operations. First is Max Pooling in which the biggest part is taken from the feature map. Next is Average Pooling that computes the average of the components in an already defined sized Image segment. The entire sum of the components in the predefined sector is calculated in Sum Pooling.

The Fully Connected (FC) Layer comprises the weights and biases along with the neurons and is used to link the neurons amongst two diverse layers. These layers are generally located former to the output layer and form the last one or two layers of a CNN structure. In this layer, the input images from the preceding layers are evened and are provided as input to the FC layer. The flattened vector then goes through a few more FC layers where the mathematical functions operations generally occur. In this layer, the classification process instigates to happen (Fig. 5).

2.2 Structure Based Types of CNNs

Depending upon their structure CNN can be classified into the following classes.

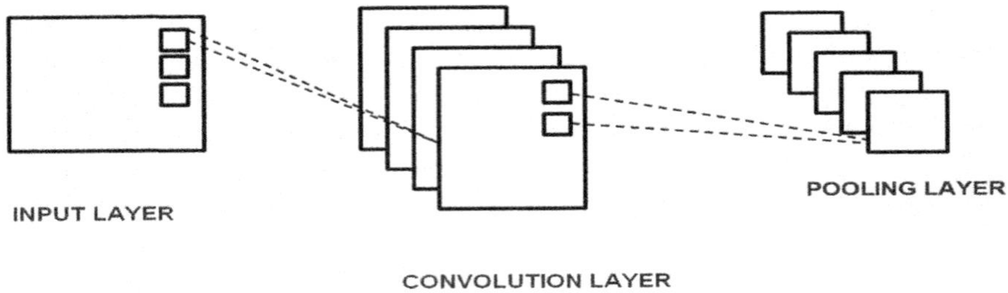

Fig. 4 Feature extraction

Fig. 5 Classification

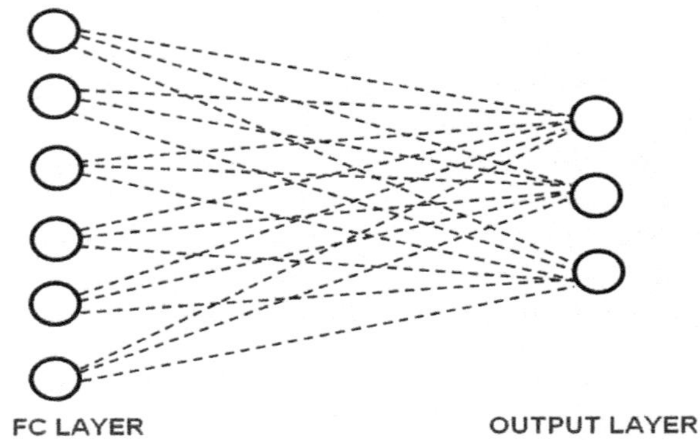

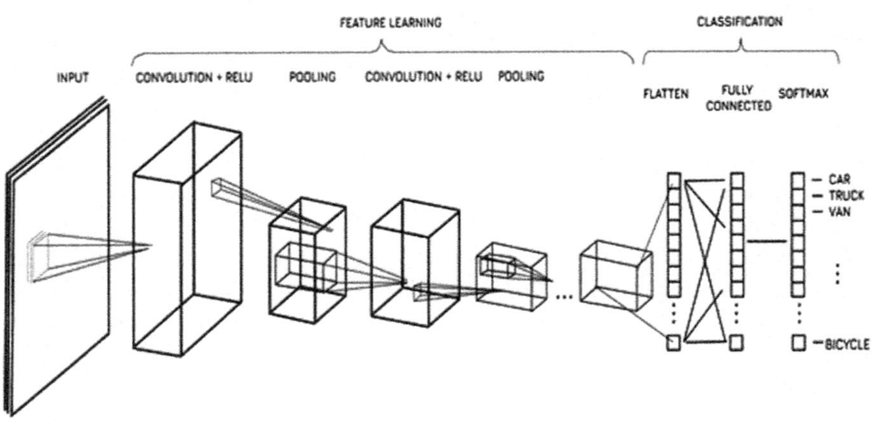

Fig. 6 Architecture of CNN

2.2.1 LeNet

LeCun and his coauthors [23] put forth a new CNN structure in the form of LeNet in the year 1989. Among the different versions of LeNet-5 (Fig. 7) is the simplest and most popular and useful. CNNs follow the feed-forward form of neural networks and the neurons comprising it are responsive to only a group of neurons surrounding them and are pretty useful in image processing in large scales. A lot of improvements have been made since the days when LeNets were introduced and this one by no means the best among the available ones nowadays. However, LeNet is the beginning of the large number of the DNNs to follow which has provided inspiration to the further developments.

Due to the non-availability of hardware support in the form of Graphics processing units (GPUs) LeNet-5 could not be popular. But, it had an electronic circuit specially designed for alteration and manipulation of memory which is responsible for the creation of images in frame buffers designated for display in an output device [8]. Also as classification models it was observed that models like SVM were outclassing LeNet at that time.

2.2.2 AlexNet

The architecture of AlexNet (Fig. 8) is similar to that of LeNet. This model won the ImageNet Large Scale Visual Recognition Challenge by employing an 8-layer in 2012 by a big margin. It broke the existing computer vision paradigm by showing that the learning of features obtained can surpass the features designed manually. The architecture provided in Fig. 7 is a revised version of the architecture of AlexNet proposed in 2012 to incorporate the fitting of two small GPUs.

In spite of the designs of LeNet and AlexNet being similar, there are a few important differences. The depth of AlexNet is higher than that of LeNet5, which is relatively small. The distribution of layers in AlexNet is into one fully connected output

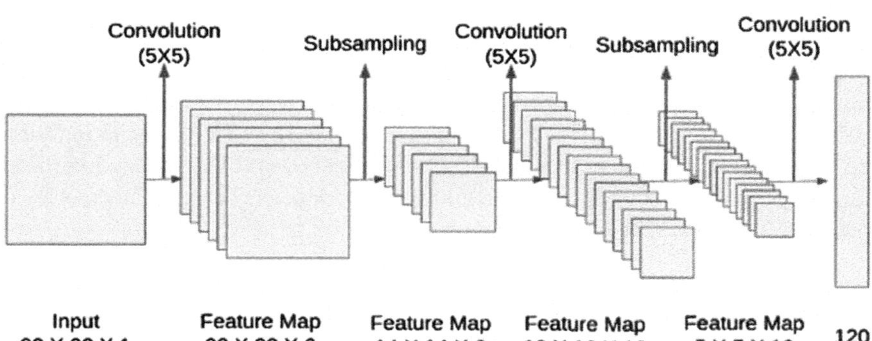

Fig. 7 Architecture of LeNet-5

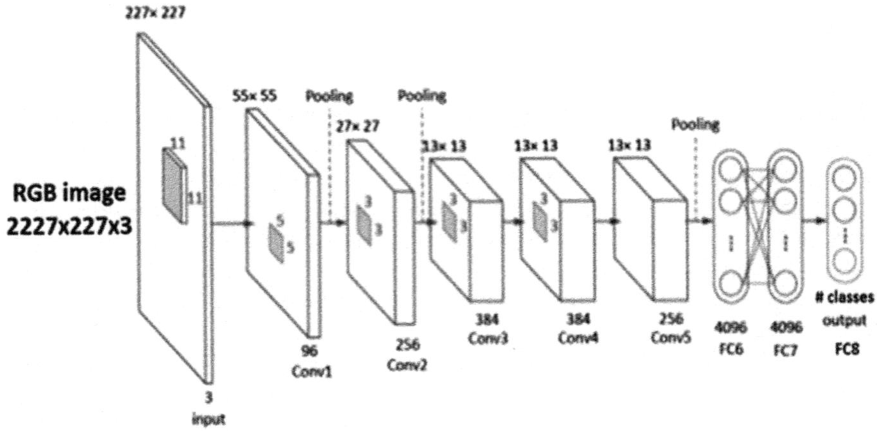

Fig. 8 Architecture of AlexNet

layer, five convolutional layers and two hidden layers which are fully connected, making a total of 8 layers. There is a difference in the use of the activation functions, while LeNet uses sigmoid activation function, AlexNet uses ReLU as the activation function.

2.2.3 VGGNet

This net has architecture having 16 convolutional layers (Fig. 9). To achieve the performance 4 GPUs are used to train for above 2 weeks and has 3×3 convolutions. The number of parameters used is very high and is fixed at 138 million. As a result the running in the inference stage becomes difficult.

2.2.4 GoogleNet

This architecture of CNN won the ImageNet Challenge in 2014 and is a large scale one. It is also called as Inception v1. Its performance is closer to that of human beings with an error rate as low as 7%. The depth of this net is 22 and the convolutions used are small being called as "inceptions" It also uses several techniques like batch normalization in order to decrease the number of parameters substantially to 40% of that used in previous models.

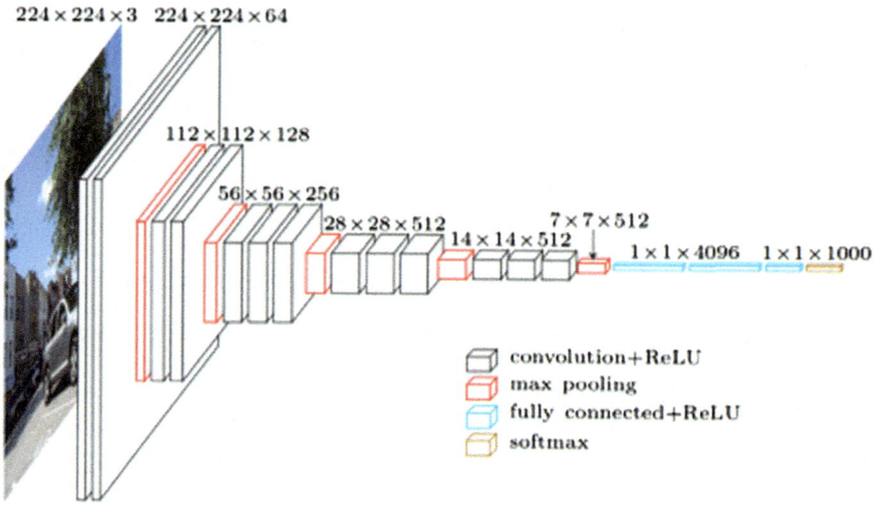

Fig. 9 Architecture of VGG Net

2.2.5 ResNet

Maximum depth used is 152 layers in the Residual Neural Network (ResNet). It uses a technique called "gated units", so as to reduce the number of convolutional layers. Use of batch normalization heavily, as is done in case of GoogleNet is helpful in running many convolutional layers simultaneously without increasing the complexity. It showed improved error rate over human level performance and won the 2015 ImageNet Challenge.

2.2.6 ZFNet

In 2013, ZFNet was invented by Rob Fergus and his PhD student M. D. Zeiler in NYU with the support from LeCun, the inventor of LeNet and comes from the surnames of the authors [42]. It is an improvement of AlexNet. The kernel used in ZFNet is of size 7 × 7 which reduces the weights significantly from the architecture where 11 × 11 size kernels are used. As a result the number of network parameters gets reduced and increases the accuracy.

2.2.7 R-CNN

The 'R' in this CNN represents "region" and it is capable of extracting objects accurately and hence the image. But, the scanning phase in this architecture takes a lot of time as it has to classify the regions. The slowness of the model is due to the algorithm used for selecting regions at the rate of 2000 per image. Later the N CNNs are used for every region and for classification it depends upon the support vector machine (SVM).

2.3 Kernel Based Types of CNN

2.3.1 1D CNN

In 1D CNN, the kernel moves in one direction. Both the input and output data of 1D CNN is 2-dimensional. Mainly used for statistical data. One-dimensional CNN can perform activity recognition tasks derived from accelerometer data, like human posture, walking, jumping, etc. This data has 2 dimensions. The primary dimension is the time step, and the other dimensions are the acceleration values on the third axiss.

2.3.2 2D CNN

In 2D CNN, the kernel moves in two directions. The input and output data of 2D CNN is three -dimensional. This is often the primary standard convolutional neural network introduced within the Lenet-5 architecture. Conv2D is typically used for image data. The general advantage of using CNN is that it can use its kernel to extract spatial features from the data, which other networks cannot. For instance, CNN can detect edges, color distribution, etc. in a picture, making these networks very powerful in image classification and other similar data containing spatial characteristics.

2.3.3 3D CNN

In 3D CNN, the kernel moves in three directions. The input/output data of 3D CNN is four-dimensional. It is usually used for 3D image data (MRI, CT scan). MRI data is widely used to examine the brain, spinal cord, internal organs, etc. Computer tomography (CT) scans also are three-dimensional data, which is an example of the creation of a series of X-ray images taken from different angles round the body. We shall use Conv3D to classify the medical data or extract features from it.

3 Application of CNN in Food Industry

Food image classification, food image processing, food image recognition are some of the common terms encountered today when we talk about applications of CNN in the food industry. The applications of CNN in food industry can be broadly classified into the following subdomains:

- Nutrient Extraction from Food Images
- Detection and Analysis of complex food matrices
- Industrial Food Packaging using CNN
- Temperature forecast using CNN in the grain storage stage
- Cuisine identification and nutrient values using CNN
- Multi-food recognition in single tray box images using CNN
- Classifying of fruits using CNN
- Computation Calories from Food using CNN
- Food Waste handling using CNN
- Food amount estimators using CNN
- Food security using CNN.

3.1 Nutrient Extraction from Food Images Using CNN

The human body needs a proper nutrient to function correctly. A balanced diet is the only form to supply nutrients to the body without making any side effects. Another advantage of consuming a balanced diet is prevention from malnutrition and health problems like diabetes, obesity, cancer, heart disease and stroke. Every user should be careful of the nutrient contents of the food they take in. The producers also should be vigilant on the nutrient contents of the food they produce, so that the customers will be able to judge it and purchase it. In fact most of the covers of the food contents display such charts. In [14], a method is developed using CNN, which can identify the food image, calculate the weight of the food and also gives the nutrient values of the food image. The block diagram of the model is as in Fig. 10.

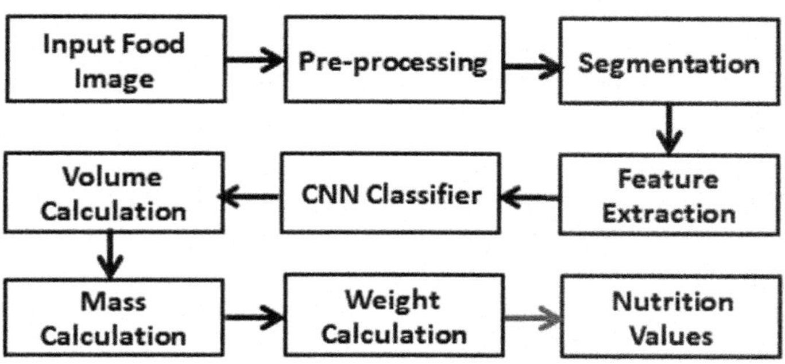

Fig. 10 CNN based nutrient extraction model

3.2 Detection and Analysis of Complex Food Matrices Using CNN

The consciousness of customers in opting for the high quality and safety food in order to improve the quality of their lives has its effect on food industries which are inclined to be able to stick to the quality of their products by detecting suitable methods for checking the food reliability and improving their quality [27]. Methods available till recently were using analytical methods for this purpose, but are destructive in nature. Some such methods are like gas chromatography, gas chromatography–mass spectrometry and high- performance liquid chromatography. These methods require very skilled operators who have to work for long hours and the processes were complex and had to enough efforts put besides using large number of chemical reagents [23]. This necessitated the development of methods which are nondestructive by nature besides being fast and accurate in food analysis and evaluation. Recent technological developments like near-infrared (NIR) spectroscopy, fluorescence spectroscopy and Raman spectroscopy the qualitative and quantitative analysis of food matrices have been improved. It has been observed that the performance of extraction of features in food, which highly depends upon data status, data acquisition has become a critical step. There are several techniques available as nondestructive techniques for data acquisition and can be used as and when required for the analyzing and detection of food attributes including computer vision ([37, 41]) and hyper spectral imaging (HSI) [10], Terahertz imaging [43] and so on. Several datasets on food images can be found [12]. These are Food-11, Food-475, UECFood-256, UECFood-100, Food-524, Food-101 and Food-5 K. In [28] discussions on recent developments in food area like fruits and vegetables, cereals and cereal products, meat and aquatic products is carried out through the introduction of 1-D, 2-D, 3-D CNN models. This approach can be further progressed with understanding of feature extraction techniques based on 1-D, 2-D, 3-D CNN for food detection and analysis.

3.3 Industrial Food Packaging Using CNN

High quality packaging and sealing is required for food materials in order to keep the food inside safe and maintain their quality. If there happens any damage to the packet in the storage place or during the transportation it will definitely lead to modifications inside the packet and hence decreasing its shelf life and the food properties are lost. The basic purpose of packaging is to protect and preserve the product and the labels on them are intended to provide information to the customer. The contamination of food and integrity of the seal during the processing and packaging are essential to decrease the food waste alongside the customer trust and belief in the context of domestic uses [15]. Before the dispatch of the packets the seals on the packets need inspection to check their status. Also, many times the quality of food is identified from the seal externally. So, there should be an inspection at every point of the route from the production centre, storage house, and transportation through standard vision systems. A context free approach with speed and accuracy are very much desirable and the acquisition time of the image and essential [40]. But, using conventional vision systems to identify the images are not suitable due to various reasons like material transparency as the public tray contains serigraphy which hides the possible faults.

Recently a lot of improved methods have been found out in computer vision which was not available a few years ago. These advancements have been possible due to the availability of various CNN models. CNNs are now applied in various areas such as speech recognition, image and video. In comparison to the traditional image classification, the top 5 methods are obtained using CNN and the development of corresponding algorithms during various ImageNet challenge competitions [35]. The development of new approaches using CNN has revolutionized this area and changed the basic philosophy. Basically, there is no necessity of feature extraction manually by consulting the subject experts and utilizing their knowledge. And the system generates these features from huge data automatically [22, 31]. As shown in [32], using deep learning approach the past few years have witnessed major performance enhancements with hyper spectral data. For classification of food products accurate use of HSI has been done perfectly [7, 11]. Similarly, the approach to determine defects in completed and raw products is developed [16]. As a result, these promising results generated by analyzing material of HSI, the CNN performance have been estimated for detecting seal contaminations on the food trays taking hyper-spectral images as data input. Computational systems are able to work in a manufacturing line have been built to determine anomaly in food trays and thus rejecting such trays during the packaging process before delivery.

3.4 Temperature Forecast Using CNN in the Grain Storage Stage

Precise prediction outcomes can heighten the consequence of ventilation choices and diminish the losses of warehoused grain. Most prevailing researches focused essentially on layer temperature forecasts whose foretell particle size is very huge. Compared to this, Ge et al. [18] used CNN-based approach that used numerous convolution layers that pooled weights to apprehend the features of grain temperature at diverse locations and made a whole use of the temperature info about the target point to predict the point temperature of grain piles. Experimentations on existent business data demonstrated that in comparison to other orthodox algorithms, CNN had the finest prediction consequence on point temperature forecast scenarios.

3.5 Cuisine Identification and Nutrient Values Using CNN

In today's fast-paced world, everyone wants information in one click. A similar rule applies once you have some food items in front of you. In social events, few cuisines are known to us while some don't seem to be. Also, in a few cases, we all know the cuisine name, but we aren't conscious of its nutritional value. This is motivational in developing a system in [33], which identified the cuisine name from the image and provided the nutrient value for it as well. Here Convolutional Neural Network (CNN) was applied to predict the cuisine name present in an image then further its nutritional value is calculated based on the information present in a database. User must click the image of the cuisine; the app identified the cuisine name and its nutrition value for normal serving amount considering the cuisine is ready using the standard recipe. The present work offered two-fold objectives of recognition of food items on plate and calculation of the nutritional value of those items. The work was pretty unique in its way for mentioned objectives as there's no such combined attempt as yet.

3.6 Multi-Food Recognition in Single Tray Box Images Using CNN

Food recognition research is growing rapidly since it can be applied in some applications, including leftover estimation. In this paper, Sari et al. [36] proposed a feature of multi-food recognition in single tray box images using Convolutional Neural Network (CNN). Automatic cropping by using rectangular contour is applied. However, for further recognition, a method to affect scarcity data using image augmentation is used. Different kind of activation function within the dense layers are tested: Rectified Linear Units (ReLU), Sigmoid, and Hyperbolic Tangent (Tanh). The result shows that ReLU activation function is that the best with no dropout and reaches 90% of accuracy. If dropout is applied in dense layer, the Sigmoid activation function is best with slightly different of accuracy 89% in recognition. It concludes that CNN is often applied to acknowledge multiple food in single image tray box.

3.7 Classifying of Fruits Using CNN

It is grounded on assessments, experiences and through scrutinizing. Harsh et al. [19] proposed a system that utilized a machine learning method to classify freshness of fruits. 2D fruit pictures were graded on shape and colour grounded investigating method. Since diverse fruits portrayals may have alike colour, size and shape values as well, so, analysis methods using colour or shape characteristics were not very operational to classify and distinguish fruits pictures. Therefore, the authors used a method to escalate the accurateness and faultlessness of the fruit freshness recognition with the assistance of size, shape and colour based technique with the amalgamation of Convolutional Neural Network (CNN). The proposed system introduced the process by snapping the fruit's picture. Then, they transferred the image to the filtration layer where the characteristics such as size, shape and colour of fruit illustrations are inhibited. Hereafter, with the use of convolutional neural network, the fruit images go through training and testing procedure. In this proposed paper, CNN is engaged to extend the size, shape and colour of fruit and with the amalgamation of these three characteristics, the attained outcomes were very promising.

3.8 Computation Calories from Food Using CNN

At the present time being overweight is one of the chief concerns all around the world. Folks treat stoutness as one of the chief illnesses since it is the core reason for so numerous health problems. The causes for obesity may be diverse such thyroid, hereditary problems, continuous medications or merely because of unhealthy eating practices and meager lifestyle. Certain people are well conscious of the fatness issue and are trying to shift to a healthy way of life. For those individuals and also for players, models and astronauts sustaining the weight are indispensable. To uphold a healthy intake, it is essential to absorb about the calorie consumption of every day or every single meal. There are numerous methods existing to measure calories, most generally from the reinstated database. It is not conceivable to measure precise calorie in the lab or it is not decent to go with the approached one. Iyanar [14] proposed a technique that measured the precise nutrient of the food by means of image processing and CNN which aids the user to compute the nutrient worth of the food even with dissimilar size or measure.

3.9 Food Waste Handling Using CNN

Nutritional assessment is perhaps the most important quality factor in a food product. Even then the salability of the product is mostly attached with the shape and the advent of goods. In [38] research was carried out with the aim to derive a suitable packaging of cherry which will decrease the waste and to enhance its ability of export and marketing betterments. In fact the basic idea involved is to use an enhanced CNN based procedure to have knowledge of the look of cherries and a classification system which is very effective. Basically, there exists two classes of cherries; irregular shaped and regular shaped. A hybrid method comprising of average pooling and max pooling with an improvised CNN method to be used from end to end after the pre-processing of the pictures in order to rank the cherries. In order to relate the proposed method (CNN) with HOG and LBP approaches, the characteristics of the images dug out by KNN, ANN, Fuzzy and Ensemble Decision Trees (EDT) procedures were pigeonholed [30]. The proposed method grounded on amalgam pooling was also paralleled with CNN with baseline pooling technique such as average pooling. Contrasts founded on the consequences of simulation established the dominance of the proposed upgraded CNN over former methods bestowing an accurateness of 99.4%. Hence, the CNN and image processing approaches are effectual in handling the marketability and exportability of the cherry fruit and can swap the outdated methods used for grading cherries at present.

3.9.1 Food Amount Estimators Using CNN

Presently, the payment procedure at eating joints is still labour-intensive and incompetent since it uses a cash inventory. A cashier has to keep track of what food is ordered, and then sum it with the cash record which is definitely not effectual. Therefore, food recognition devices and programmed food amount estimations have the response to these shortages which is proposed in [34]. Food detection intends to simplify expense at eateries, and automate food bill approximation using the Convolutional Neural Network (CNN) classification technique. The recognition accurateness of six kinds of food using this proposed CNN technique was gained 100% with 0.8% data partition training data and 0.2% test data with 9000 epochs and learning rate of 0.0002, with a recognition time of less than 10 s.

3.9.2 Food Security Using CNN

An important topic in food science and monitoring the quality and maturity is the testing of agricultural products through harmless testing techniques. This is specifically designed to determine the chemical properties of using traditional mathematical methods. However, these mathematical methods do not provide sufficient information as they do not reflect difficulties when working with natural products. While in-depth study gained higher acceptance than state-of-the-art art estimates, has only found interest in the use of harmless experiments over the years. As a result, height in the interest of in-depth education can be reverted to learning data representation by extracting features internally hidden space that cannot be determined precisely. Food security is promoted as a goal of precise and smart horticulture when product quality and yield extension is achieved by reducing the application of pesticide simultaneously. In [29] detection of apple scab is performed in the early stage of development, using CNN and mobile phones. In [21] the research carried out using CNN training had an acquisition of two datasets are collected which include scab attacked fruit images and leaves from an apple tree. Since the probability factor is involved in scab appearance and acquisition of data may take much time transfer learning is used as an alternative methodology. In this research the aim was to use the topmost dataset which is suitable and apply transfer learning for the purpose of apple scab detection. Also, the effect of transfer learning is adjudged through a comparison of learning from the scratch. The statistical analysis confirmed the positive effect of transfer learning on CNN performance with a significance level of 0.05. A comparison of the results obtained with the results obtained in related researches shows that if the imbalanced data problem can be resolved then the accuracy of application is to be enhanced.

4 Future Trends of CNN in Food Industry

CNN has enough potential as is seen from the discussions above. We outline some of the future trends in the application of CNN further in coming days.

4.1 Material Identification

CNN can be used in identifying polymers, plastics, and micro plastics automatically using computer vision to save hundreds of hours in sorting and recycling.

4.2 Product Development System

CNN can also be used to identify polymers or plastics most suitable for new products to save hundreds of hours in research and development.

4.3 Packaging Recommendation System

CNN models can be developed to identify packaging options that suit best for new products using recommendation systems. This can help in saving hundreds of hours in product development.

4.4 Hygiene in Food Plants

CNN has been utilized in estimating hygiene in kitchens. The same procedure can be applied to promote better hygiene in a food plant which is equally important. It is imperative that there must be surveillance to ensure compliance with regulations. The system, which may even be utilized in restaurants, uses cameras to watch workers, and it uses facial-recognition and object-recognition software to work out if workers are wearing hats and masks as needed by food safety laws. If violations occur then the system can extract screen images for review.

4.5 Cleaning of Wash Equipment

Washing the equipment using advanced techniques definitely helps in improving standards in health. Efforts by researchers in University of Nottingham, in which they have developed a new system uses AI techniques which reduces the time in cleaning, reduces the use of resources which is saving almost £100 million ($128.8 million) in a year. It is termed as Self-Optimizing-Clean-In-Place (SOCIP), and uses optical florescence imaging with ultrasonic sensing so that the images of food residue can be detected to the minutest level. It is felt that for optimizing the process of cleaning a CNN based system can be developed.

4.6 Predicting Crop Yield

Crop production depends on a number of factors like humidity, soil type, temperature, etc. Predicting crop yield is a tiresome job as this is not precise data, and predictions might not always turn out to be right. There are certain disadvantages using deep learning as it cannot handle uncertain and imprecise data, but under these circumstances fuzzy logic can be used to have good effects [39]. It is more reliable for real world data. Fuzzy Logic complements this drawback of DCNN. It can help in crop yield prediction to overcome the drawback of data imprecision and get better accuracy in much less time. This can be one of the major trends of CNN related to food production in the future.

4.7 Improving Quality of Farm Produce

CNN-NLP research can be conducted beyond just normal quality testing of products. A survey can be done state wise where every civilian can participate. Government can get proper insights about the quality of the farm produce in a state and improvise it by providing farmers with better quality pesticides or seeds. Farmers on a personal level can adopt healthy practices to promote food safety.

5 Conclusions

In this chapter, we started in making an extensive study on applications of CNN in food industry. We introduced the different types of DNNs, CNNs and their evolution elaborately. CNN being a classification model it is suitable for processing of images. Deep Learning (DL) has Grounded on this pronounced attention that CNNs have had in the preceding years, we present a review of the use of CNN applied to diverse applications of CNN in the food industry. Our points of study were Nutrient Extraction from Food Images, Detection and Analysis of complex food matrices, Industrial Food Packaging using CNN, Temperature forecast using CNN in the grain storage stage, Cuisine identification and nutrient values using CNN, Multi-food recognition in single tray box images using CNN, Classifying of fruits using CNN, Computation Calories from Food using CNN, Food Waste handling using CNN, Food amount estimators using CNN and Food security using CNN.

We collected several directions of further research using different forms of CNN in food industry. It is expected that enhanced research on the topics presented and directions provided will improve the application area of CNN in food industry.

References

1. A. Adate, B.K. Tripathy, Understanding single image super-resolution techniques with generative adversarial networks, ed. by J. Bansal, K. Das, A. Nagar, K. Deep, A. Ojha. Soft Computing for Problem Solving. Advances in Intelligent Systems and Computing, vol 816 (Springer, Singapore, 2017), pp. 833–840
2. A. Adate, B.K. Tripathy, Deep learning techniques for image processing, in *Machine learning for big data analysis*. ed. by S. Bhattacharyya, H. Bhaumik, A. Mukherjee, S. De (De Gruyter, Berlin, Boston, 2018), pp. 69–90
3. A. Adate, B.K. Tripathy, S-LSTM-GAN: shared recurrent neural networks with adversarial training, ed. by. A. Kulkarni, S. Satapathy, T. Kang, A. Kashan. Proceedings of the 2nd International Conference on Data Engineering and Communication Technology. Advances in Intelligent Systems and Computing, Vol 828 (Springer, Singapore, 2019), pp. 107–115
4. A. Adate, B.K. Tripathy, D. Arya, A. Shaha, Impact of deep neural learning on artificial intelligence research, ed. by. S. Bhattacharyya, A.E. Hassanian, S. Saha, B.K. Tripathy. Deep Learning Research and Applications (De Gruyter Publications, 2020), pp. 69–84
5. A. Adate, B.K. Tripathy, A survey on deep learning methodologies of recent applications, ed. by. D.P. Acharjya, A. Mitra, N. Zaman. Deep learning in data analytics- recent techniques, practices and applications (Springer Publications, 2021), pp. 145–170
6. C.M. Agbai, Application of artificial intelligence (AI) in the food industry. GSC Biol. Pharm. Sci. **13**, 171–178 (2020)
7. M. Al-Sarayreh, M.M. Reis, W.Q. Yan, R. Klette, Deep spectral-spatial features of snapshot hyper spectral images for red-meat classification, In 2018 international conference on image and vision computing New Zealand (IVCNZ), (2018), pp. 1–6
8. L. Alzubaidi, J. Zhang, A.J. Humaidi, A. Al-Dujaili, Y. Duan, O. Al-Shamma, L. Farhan, Review of deep learning: concepts CNN architectures, challenges, applications, future directions. J. Big Data **8**(1), 1–74 (2021)
9. S. Bhattacharyya, V. Snasel, A.E. Hassanian, S. Saha, B.K. Tripathy, *Deep Learning Research with Engineering Applications* (De Gruyter Publications, 2020)
10. W. Cheng, D.W. Sun, H. Pu, Y. Liu, Integration of spectral and textural data for enhancing hyperspectral prediction of K-value in pork meat. LWT Food Sci. Technol. **72**, 322–329 (2016)
11. J.-H. Cheng, B. Nicolai, D.-W. Sun, Hyper spectral imaging with multivariate analysis for technological parameters prediction and classification of muscle foods: A review. Meat Sci. **123**, 182–191 (2017)
12. W. Ciocca, D.W. Sun, H. Pu, Q. Wei, Hetereospectral two-dimensional correlation analysis with near infrared hyperspectral imaging for monitoring oxidative damage of pork myofribrils during frozen storage. Food Chem. **248**, 119–127 (2019)

13. R. Debgupta, B.B. Chaudhuri, B.K. Tripathy, A wide ResNet-based approach for age and gender estimation in face images, ed. by. A. Khanna, D. Gupta, S. Bhattacharyya, V. Snasel, J. Platos, A. Hassanien, *International Conference on Innovative Computing and Communications, Advances in Intelligent Systems and Computing*, vol 1087 (Springer, Singapore, 2020), pp. 517–530. https://doi.org/10.1007/978-981-15-1286-5_44
14. D.P. Dharani, D. Iyanar, CNN based nutrient extraction from food images. In 2020 Fourth International Conference on I-SMAC (IoT in Social, Mobile, Analytics and Cloud)(I-SMAC) (IEEE, 2020), pp. 398–402
15. M. Dudbridge, *Handbook of Seal Integrity in the Food Industry* (Wiley, 2016)
16. C.H. Feng, Y. Makino, S. Oshita, J.F.G. Martín], Hyper spectral imaging and multispectral imaging as the novel techniques for detecting defects in raw and processed meat products: Current state-of-the-art research advances. Food Control, **84**, 165–176 2018. http://www.sciencedirect.com/science/article/pii/S0956713, 517303602
17. Y. Freund, D. Haussler, *Unsupervised Learning of Distributions on Binary Vectors Using Two Layer Networks, Technical Report UCSCCRL94-25* (University of California, Santa Cruz, 1994)
18. L. Ge, C. Chen, Y. Li, T. Mo, W. Li, A CNN-based temperature prediction approach for grain storage. Int. J. Internet Manuf. Serv. **7**(4), 345–357 (2020)
19. A. Harsh, K.K. Jha, S. Srivastava, A. Raj, S. Raghav, Fruit freshness detection using CNN approach, twitter, our grim reality (2021)
20. https://becominghuman.ai/introduction-to-artificial-intelligence-5fba0148ec99
21. S. Kodors, G. Lacis, O. Sokolova, V. Zhukovs, I. Apeinans, T. Bartulsons, Apple scab detection using CNN and transfer learning. Agron. Res. **19**(2), 507–519 (2021)
22. G. Lacey, G.W. Taylor, S. Areibi, Deep learning on fpgas: past, present, and future (2016). arXiv preprint arXiv:1602.04283
23. Y. LeCun, B. Boser, J.S. Denker, D. Henderson, R.E. Howard, W. Hubbard, L.D. Jackel, Backpropagation applied to handwritten zip code recognition. Neural Comput. **1**(4), 541–551 (1989)
24. Y. LeCun, Y. Bengio, G. Hinton, Deep Learning. Nature **521**, 436–444 (2015)
25. G.E. Hinton, S. Osindero, Y. The, A fast learning algorithm for deep belief nets. Neural Comput. **18**, 15271554 (2006)
26. Z. Li, W. Yang, S. Peng, F. Liu, A survey of convolutional neural networks: analysis, applications, and prospects (2020). arXiv preprint arXiv:2004.02806
27. Y. Liu, H. Pu, D.-W. Sun, Hyperspectral imaging technique for evaluating food quality and safety during various processes: a review of recent applications. Trends Food Sci. Technol. **69**, 25–35 (2017)
28. Y. Liu, H. Pu, D.W. Sun, Efficient extraction of deep image features using convolutional neural network (CNN) for applications in detecting and analyzing complex food matrices. Trends Food Sci. Technol. **113**, 193–204 (2021)
29. K. Maheswari, A. Shaha, D. Arya, B.K. Tripathy, R. Rajkumar, Convolutional neural networks: a bottom-up approach, eds. by. S. Bhattacharyya, A. E. Hassanian, S. Saha, B.K. Tripathy, Deep Learning Research with Engineering Applications (De Gruyter Publications, 2020), pp. 21–50
30. M. Momeny, A. Jahanbakhshi, K. Jafarnezhad, Y.D. Zhang, Accurate classification of cherry fruit using deep CNN based on hybrid pooling approach. Postharvest Biol. Technol. **166**, 111204 (2020)
31. M.E. Paoletti, J.M. Haut, R. Fernandez-Beltran, J. Plaza, A.J. Plaza, F. Pla, Deep pyramidal residual networks for spectral–spatial hyperspectral image classification. IEEE Trans. Geosci. Remote Sens. **57**(2), 740–754 (2019)
32. M. Paoletti, J. Haut, J. Plaza, A. Plaza, Deep learning classifiers for hyper spectral imaging: a review. ISPRS J. Photogramm. Remote. Sens. **158**, 279–317 (2019)
33. D. Pawadea, A. Dalvia, I. Siddavatamb, M. Carvalhoc, P. Kotianc, H. Georgec, Cuisine detection using the convolutional neural network. Int. J. Educ. Manage. Eng. **10**(3), 1 (2020)
34. A. Ramdani, A. Virgono, C. Setianingsih, *Food detection with image processing using convolutional neural network (CNN) method*, In 2020 IEEE International Conference on Industry 4.0, Artificial Intelligence, and Communications Technology (IAICT), IEEE, 2020, pp. 91–96

35. O. Russakovsky, J. Deng, H. Su, J. Krause, S. Satheesh, S. Ma, Z. Huang, A. Karpathy, A. Khosla, M. Bernstein, A.C. Berg, L. Fei-Fei, ImageNet large scale visual recognition challenge. Int. J. Comput. Vision **115**(3), 211–252 (2015). https://doi.org/10.1007/s11263-015-0816-y
36. Y.A. Sari, S. Adinugroho, J.M. Maligan, M.A. Rahman, Y.G. Bihanda, Multi-food recognition in single tray box image with scarcity data using convolutional neural network, IICST 2020, 5th International workshop on Innovations in Information and Communication science and technology (Malang, Indonesia, 2020), pp. 63–69
37. D.W. Sun, T. Brosnan, Piza quality evaluation using computer vision, part2- pizza topping analysis, jour. Food Eng. **57**(1), 91–95 (2003)
38. G. Tavill, Industry challenges and approaches to food waste. Physiol. Behav. **223**, 112993 (2020)
39. B.K. Tripathy, J. Anuradha, *Soft Computing- Advances and Applications* (Cengage Learning publications, 2015)
40. P. Vithu, J. Moses, Machine vision system for food grain quality evaluation: a review. Trends Food Sci. Technol., **56**, 13–20. http://www.sciencedirect.com/science/article/pii/S092422441630084X
41. H.H. Wong, D.W. Sun, Assessment of cheese browning affected by backing conditions using computer vision. J. Food Eng. **56**(4), 339–345 (2003)
42. M.D. Zeiler, R. Fergus, Visualizing and understanding convolutional networks, eds. by. D. Fleet et al., ECCV 2014, Part I, LNCS 8689, pp. 818–833 (2014)
43. L. Zhou, C. Zhang, F. Liu, Z. Qiu, Y. He, Application of deep learning in food: a review. Compr. Rev. Food Sci. Food Saf. **18**(6), 1793–1811 (2019)

Utilization of Artificial Intelligence in the Food Industry

Vikas Garg and Neha Puri

Abstract The introduction of Artificial Intelligence (AI) in the food industry streamlines the quality assessment of food products. The innovation has brought many advanced applications to make food fabrication more proficient, safe, and gainful. For example, advances in hereditary design are prompting the creation of harvests and food items. The items are more advantageous for the purchasers and more effectively cultivable for the makers. The five artificial intelligence application categories include sorting products and packages, Food safety compliance, Improved cleanliness, product development, and marketing. Artificial Intelligence has directly or indirectly impacted the food industry. The basic food item industry utilizes AI to offer clients focused on offers, oversee stock, and decrease squander. Artificial intelligence applications will impact the food handling organizations and help them build their income and lift client experience. Furthermore, numerous food-centered AI stages are accessible to shoppers. For instance, Softwear like Wellio utilizes AI and conduct science to give customized formula proposals and afterward allows clients to arrange their goods on the web. Habit creates customized nourishment plans dependent on the consequences of a sustenance test. Food processing is a convoluted business and it includes arranging the food or crude materials coming from the homestead, keeping up the hardware and a few kinds of gear, and then some. However, artificial intelligence will be a boon for this industry and will make the work more tranquil and smoother.

Keywords Artificial intelligence · Purchaser · Applications and food product

V. Garg (✉) · N. Puri
Amity University, Noida, Uttar Pradesh, India
e-mail: vgarg@gn.amity.edu

N. Puri
e-mail: npuri@amity.edu

A. E. Hassanien and M. Soliman (eds.), *Artificial Intelligence: A Real Opportunity in the Food Industry*, Studies in Computational Intelligence 1000,
https://doi.org/10.1007/978-3-031-13702-0_6

1 Introduction

In every country, Agriculture sector plays a very vital role. However, the sector is facing problems in terms of lower productivity and less profit margins. The production unit is sinking because smaller companies cannot predict the choice of the customers or cannot comply with the food regulations. That's the main reason for incorporating the need of Artificial Intelligence in the food industry [5, 8, 9, 13, 14].

In such a case, Artificial Intelligence is our only hope. It is an advanced mechanism which can do a lot of work. Large companies are employing AI technology to increase their efficiency and profits [7]. Companies like Coca-Cola and Kellogg's [22] use the technical mechanism and applications to promote rare fusions. The adoption of AI in the food industry will increase the market reach, lead to cost-saving, enhance the guest experience and reduce errors as shown in Fig. 1.

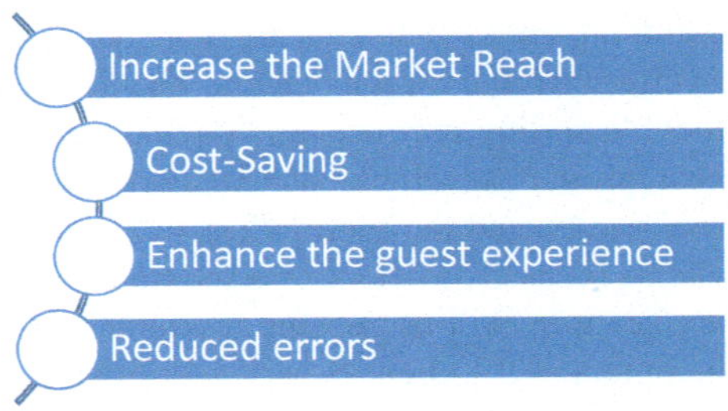

Fig. 1 The adoption of AI in the food industry

The support mechanism behind the agricultural industry is food manufacturing. It entails categorizing livestock and final production review. Under special situations, the unit must hold the machinery and storage. In addition, it must distinguish between specific items and find multiple ways of storing them. Output will depend on the choice of the client and the availability of capital. Taste curators are actually battling teeth to teeth to create fantastic food platforms and gain immense likes.

Artificial Intelligence is changing lifestyles. From construction industries to homes we are using AI to ease our work. The question is how it will help the food industry? According to makers, AI can predict weather forecasting and alert the farmers about incoming storms. It can also help the goods transport to improve yields and impact costs. The food industry depends on a lot of timing and temperature. If AI is implied, sorting, processing, packing will become more feasible. Moreover, the company can keep a close eye on the producers using robots and drones.

2 Overview of AI in Food Industry and Need of AI in Food Industry

Food and Handling Industry (FP&H) and its continuous involvement with artificial intelligence (AI) and technology is not the new thing we are seeing. Finding food and getting it whenever comes to mind through applications and online is becoming common nowadays. We all are seeing the food are prepared and packed by the robots and ordering taking a few places and outlets are also done through AI and robotics [2, 6, 11, 20]. Technological advancements and tech development also helps the process becoming faster like packaging and overall food safety. Artificial Intelligence (AI) its advancement not only in food industry but also catching attention from all domains of business-like technology to corporate houses, AI is controlling everywhere. Nowadays we cannot even imagine industries running without having good or minimum knowledge of AI. The food processing and handling industry (FP&H) will continue to grow at a very speedily rate till 2021 after that too and the reason is AI. The actual industry (FP&H) is running at a whopping above $100 billion and will continue to expand in coming decades.

The job opportunities creating due to AI and especially in the food processing and handling industry are at high rate as to maintain the technology even you need a high skilled and low skill people and packaging and handling almost require human intervention with great contribution of AI and technology to it making the complex things into very simpler for which human require lots and lots of efforts and time, but with the usage of AI and technology to the (FP&H) is making the industry growing at an unimaginable rates and will continue to do so in the future as well. The overall benefit which the sector will get from AI is increase in their revenues with improving the standards of food while cutting the cost.

According to one report published by McKinsey [20], which says that the (FP&H) sector will grow at a constant rate with some ups and downs around 4–5% at least till 2021. Artificial intelligence will prove to be a revolutionary move if applied carefully seeing all pros and cons. Cost management, improving the standards of food and overall industrial development will take place using AI in the sector.

Seeing the overall benefits and importance of AI to (FP&H) Sector, here are some of the needs of Artificial Intelligence (AI) in the food Industry.

1. Supply chain management
2. Maintenance alert and monitoring
3. Customer preferences
4. New product line
5. Food delivering via online platforms.

Artificial intelligence and its wide horizon in terms of usage and work in food industry is not only taking industry towards profitability but also it is creating new opportunities for employment of various specialist like data analyst, AI specialist, technology driven people.

3 Need of AI in the Food Industry

From supply chain management, maintenance, customer preference, introducing to new products and food delivering. No doubt food packaging and handling is a complicated process which require attention at each and every step to avoid any external contamination get involve with it [14, 15].

Several processes are involved and many are interlinked with AI and technology these things are done at a very easily. Human intervention with technology makes the product more cheaply, better quality and with less time making the industry to produce more and earn more profits.

However, there are many FP&H industries where the complete process is driven by AI and technology and human intervention is there. Technology is a vital component in connecting all the things human with AI and AI with the food industry. Below are the 5 important uses of AI that will benefit and are benefiting the food industry with better supply management and touching the revenues to new heights.

1. **Supply Chain Management**

Management of supply chain is very crucial stage for any product especially for food products because it needs to be delivering on time and reach to customers on time. Supply chain and its introduction with AI is taking it to the new levels. Supply chain management also becomes important part of any industry because without any good and systematic supply chain of products and services industry cannot grow out. Its transportation and proper carriage is needed to deliver it to its destinations from the origin. Artificial intelligence (AI) enables this system to move in a very systematic and flawless manner. Everything is being tracked with the help of artificial intelligence enablers systems and software. Software enables all those things manageable which were not applicable in the traditional way of supply chain management. Consumers and suppliers are now connected to one link of AI and all are benefiting from it. Customers are getting their products deliver on time and suppliers and manufacturers are earning more and more. AI is enabling all these processes to get it done on time and before time with generating more revenues. Simplified processes with higher output and cost management during the production makes things easier for food industry. The best thing about supply chain management enabling with artificial intelligence (AI) is integrating all other sector with it within the efficiency and proper limits. Exploring of AI with supply chain is not a very new concept. It is being from last many years and constant changes are being made with new technological advancements and best methods to introduce. Traditionally, supply chain management was moving at its own pace and not on the pace which needs to be there. All the things were dependent on the transportation and how the transporter takes the things and reaches to the desired destination. It was not decided on many product line that amount of time will it take to deliver because there is nothing to track what transporter is doing with the products and how they are dealing with it. Strong connectivity and a proper reach needs to be there between all the coming segments of supply chain for proper Management. Enabling AI with supply chain not only enables the efficiency but also enables the proper tracking of the product dispatching from its actually place to reaching to the desired location.

Supply chain integrating with AI has multiple and diverse range of benefits from manufacturers to suppliers and finally to the local markets reaching to the customers. All are now connected with a single string of AI. Customer can track its own order if placed via online and that too comes under supply chain management of the food and other veggies. From tracking of the orders to online payments all is possible just enabling of AI with the food industry [14, 15].

For example: Swiggy and Zomoto [21] are food ordering websites and all their technologies are connected with AI. With the help of AI, they are getting to know what their consumer wants, what age people eat what kind of food and consumer preference is all the companies want to know and grow with that. Getting to know about consumers and their preferences is the most difficult task company feels but after enabling AI and integrating it with the software all becomes easy and the entire data just one clicks away. Supply chain and its Management is enabling all the sectors interconnected with food industry are growing with rapid process as it well known fact that food industry is the industry in which recession will never come, downfall will never be seen because people will eat what they want to and if AI is giving more options the chances are less to avoid.

2. **Maintenance Alert and Monitoring**

Food industry manufacture the food product at a very large scale and to produce that on such large scale complex mechanism is being followed. Though all these mechanisms are not followed and cannot be followed by using human intervention and at these complex situational points where humans can work up to some extent and after that a strong mechanical force and strength is needed. AI enabling software and robotics machines enabling these entire things at a very short span of time and without any human intervention things get done and with best and possible effectiveness it can bring out [2, 6, 11]. Depending upon the kind of work machines are performing their maintenance too required that kind organization. More the complex machine will be, more efforts need to be made to get it correct and proper working as it was earlier. Machines are always being the best inventions of mankind solving and easing all the tough work and situations where human interventions are not at all possible. Enabling AI integration software to the robotics machinery and other tech devices made the work easier with helping is cost cutting of the organization. These AI enabled monitoring and a maintenance system helps the organizations to get an alert for the maintenance and service of the machinery. Typically, what generally these systems do is that the program is already being installed in them and after specific period of time and based on how much the amount of machine is working it gives an alert for the maintenance and due to these enablers and fast rapid AI based system organization work is reaching to the efficiency because they are well aware of their machinery and when it next required its maintenance. It helps both in boosting the efficiency of the food and manufacturing unit and also gives rise in the revenue of the organization. Artificial intelligence to the modern technology will going to transform the sector way out and how it used to perform. All the capabilities that the industry is looking for in the artificial intelligence and how it going to maintain the systematic of machinery and its timely maintenance record too. Data science and its correlation with artificial intelligence is also booming the food industry with its important features that many big food brands and organizations are adopting to its system and their work style totally on it. AI and its integration with the food industry not only benefits the organization's linked to food and other segments but also to the consumers and buyers as they will be getting more options and preference's. Machine learning [18, 19] and its analysis is also the new trend which is coming up and allows the food manufacturing to ascertain the quality and the effectiveness of the machine and how well it is performing. The total quality check and the performance check is completely dependent upon the AI based machine learning software and its interlinked devices. After getting the overall quality check of the machine, which basically tells the organization that up to which level the machine can work and what is its actual effectiveness of work. This type of program is really a great segment in the manufacturing sector where effectiveness is always at the peak and you want to give the best input to obtain the quality and desired product. Data analytics and its interpretations helps the organizations to eliminate the problem at its very beginning stage and later on it will not hamper the product and its making. Food industry is filled with so much big giants and restaurant's chain that you need to have a very specified and clear product line to achieve more from the market in terms of higher revenues and profits.

3. **Customer Preference**

Matching with the needs that what your consumer wants and what type of segment they are more interested in. food industry is said to have a booming and bright future in the coming decades because taste is what all matters in the food industry and consumer will never ever leaves the taste. From benefitting to food processing unit, AI also enables customer support and letting them know with their decision making for better purchase and buying. AI in decision making of customer is really an advancement to give a new experience to the customer, a new way of shopping. Shopping with your decision and a way you want to shop. This is what AI teaches customer and big brands. AI has enabled the various options and features in its graphic to enable customer a new way of shopping experience. For example: Kellogg's [22] has introduced the AI support customer support with the help of IBM software's and technologies which enables customer to mix ingredients of 2 different kinds and letting them know whether they will go with each other or not. Data science and data analytics is the new advancements in every field and with its new and new exploring in food industry everything is shaping into the new era and everything is transforming to the technological world. There are many examples in the food industry world that are using data science and analytics to know the consumer preference and what they actually want. Through online delivering food options restaurant's get to know actually how consumers are reacting to new segment of food line and able to generate their taste through data analytics. The Coca Cola enabled the fountain machine creating method for this using AI enabled technologies and software's which used for creating for a new soft drink by customer themselves. They just have to mix up the two combinations of flavours like berries and cherries and based upon these choices Coca Cola decides to launch which of the best drinks in the market. These types of processes are very complicated if we involving human interventions to it but after installing of fountain drink combination machine across USA with enabling of AI and its automation self-driven technology, everything becomes very easy and reliable and the company the get the data of whatever they need whether it is in terms of new addition floured drinks, customer data, customer age using that fountain machine and which age group is preferring which type of flavour, all these important and relevant question came into existence only and only when there is AI, IoT and Automation self-driven technologies used. With using up of fountain drink machine Coca Cola came to know that majority of the population has mixed cherries with the sprite and later on with this data they invented and launched Cherry Sprite into the market and that too was a huge success because it was already known and developed through and by customers only, what they did is just collecting the data with the help of AI algorithm's and launched it. Many food ordering companies promise to deliver the food within a particular time limit no matter what your location is, this all is happening because of the data analysis and learning of algorithms. Food ordering through that analysis makes a segment of each delivery site and how watch thing needs to be executed within the time limit as promised.AI is not only the change maker to the food industry but with that integration of data analysis and algorithms all became possible to reach out to the desired segment of customers. Artificial intelligence and its link to consumer preference's is actually making the competition tough in the market because now the time has come that every new industry starts from consumer preference's through various data analysis and research.

4. **New Product Line**

As we are moving towards industry 4.0, which actually means everything needs to automated, self-driven and totally work should be done through machines and humans efforts and work should have turned out rom this to R&D for more development of these of automation and self- driven machines. Food industry is going through several changes and with that new products are launching day by day to attract the customers, to retain the customers and most important to increase the revenue and growth of the organization. The (FP&H) sector is itself very unique and every time it needs innovation to satisfy the customer with something new products, but the question is how the companies get to know what to develop? And what kind of product needs to be launched in the market. Every time there must be something new, something connecting to the consumer day to day needs then only a product will up stand in the market and between the competitors. Before launching of AI, automation and IOT the companies needs to have campaign advertisement to check the taste of the consumer and what they actually want. The interesting segment which developed was that the mixing and making it into a new form. The intelligent science and analysis is under process by the scientist which actually enables the cook to smell the new ingredient and make it into the new variety and add on to the menu as a new food. These type of changes are required are we saying and talking all about industry 4.0 then why not this 4.0 starts with the food industry itself which has the no phase of recession only consumer preferences and taste will vary time to time and industry needs to adapt with that taste to new introduce new product line to have that customer attract towards them. The very interesting aspect of this intelligent science is not only that the restaurant's will be able to generate new types of cuisines but they will also lead the competition in the industry because if you have to survive you have to continuously and gradually change and adapt to the new technological advancement's happening around and implementing it. Cooking by the chefs according the mixing and choice of customer will be going to be the trend. The time is not so far where not only dishes but type of ingredient preference would also be there based upon AI and algorithms and to check which two or three type of combination of ingredients will go with each other and which type of dishes will choose for it. Artificial intelligence and the preference of the food will be going to change the look of food industry in the coming decades and future. Intelligent science and its advancement not only limit to this but through this restaurants chain and big hotel group are trying to find out that what kind of mixing of ingredient's will bring out which dish and how regional wise targeting and to selected group of people and age what impact will come. Intelligent science with algorithms can be a change maker to the food industry but these types of algorithms are most important in deciding what to launch next and what kind of flavour to launch in the market. This type of logarithmic system requires lots of efforts to interpret the data but after enabling of AI these interpretation becomes so easy that launching of products in the market. Around the globe there are many example of American and western food which generally does have the same mixing of flavours which is generally done not by Asian dishes and the south Indian dishes. In order to find this change and why it is happening, intelligent science and data analytics is very useful to determine what consumer are wanting within the same flavour range. New innovations are there in industry and with the more usage of AI and IOT the industry will transform its style and way of presenting.

5. **Food Delivering via Online Platforms**

AI, IOT and automation has transformed the processes and carrying out style of products which has also benefitted the organization to speed up their profits and revenues. Changes are difficult with AI for small scale manufacturing units and food processing units where in 2020 too majority of the work is being done by the labour and under their supervision too. People are relying upon AI and automation, adapting to their technologies and style but still it will take a long way to achieve complete automation in industries. It is not wonder now that you think of the food and it's just one click away from your thinking. Everything is automated and digitized. What new restaurant's and hotel chain are doing they are trying to give the most advantage to the customers while sitting at home and eating at their own comfort. Food delivering options are many in today's just look for it and there and option in front of you. Reinventing food delivery with using all the algorithms and data science companies used to find out that how to deliver the best and the tasty food at your door step. It's all about artificial intelligence, what you think, what you choose and what you want is all now captured by the companies in the form of data using analysis, online surveys and all these complex algorithms are converted to turn out be a new product line for the consumer because it is them who tells the company what to launch next and how to launch. Through AI companies are not making high revenues but they are also cost cutting there some unnecessary costs which were not possible to do with traditional way of Management. Consumers just have to think for a like Swiggy, Zomoto, Uber eats these all are what, and they all are providing the food to the customers reaching out to their door step. AI and its usage to the delivery based systems is new but still not very new because tracking of your orders and live location can easily be done through to it. AI and other means of it cannot be overall change maker if industries and companies are not trying to change themselves then best to best technology will degrade with time. The online food ordering services is not only expanding in India and Asia but to all parts of the world. It has now become the necessity for the common people to order food via online and get it deliver at doorstep. Every food online ordering platform has the large number of information and client data and through machine learning and other algorithms to check the consumer taste and its preferences. Market domination is still one of the important aspects of the food industry where through data science and intelligence network of logarithmic companies are having large number of data but that data needs to have a strong direction. Reallocating those resources and putting back those to greater experience with AI and robotics experiencing the better customer satisfaction. Machine learning can be an advancement to these AI by developing major changes to these things. Data science can prove to be challenging to implement by the companies and thus further complication linked to it can be a reason to think but over and all and gradually the AI and its technological advancements will benefit the customer and that to by providing new way of customer experience.

Key drivers of Artificial Intelligence in Food Industry

There are five key areas where Artificial intelligence has directly impact on food industry [17] as shown in Fig. 2.

1. Predictive Quality and Yield—For producers of all kinds, reducing manufacturing losses and avoiding production process inefficiencies has always been a relentless challenge. Today, as increasing demand encounters intensified rivalry, this is truer than ever. In the one hand, the demands of consumers are greater than ever before; global market preferences are increasingly "westernizing," even though the demographic boom persists. In this article, with the use of machine learning techniques and Edge Cloud Computing technologies, we discuss a new integrated approach for predictive model-based quality inspection in industrial development [1]. In order to expose the latent sources of many of the perennial output losses farmers face on a regular basis, Predictive Quality and Yield uses Industrial Artificial Intelligence. Using Machine Learning algorithms that are specially qualified to grasp each individual development process intimately, this is achieved by continuous, multivariate analysis. "Supervised learning" is the specific AI/Machine Learning technique used here, where the algorithm is trained to identify trends and patterns.

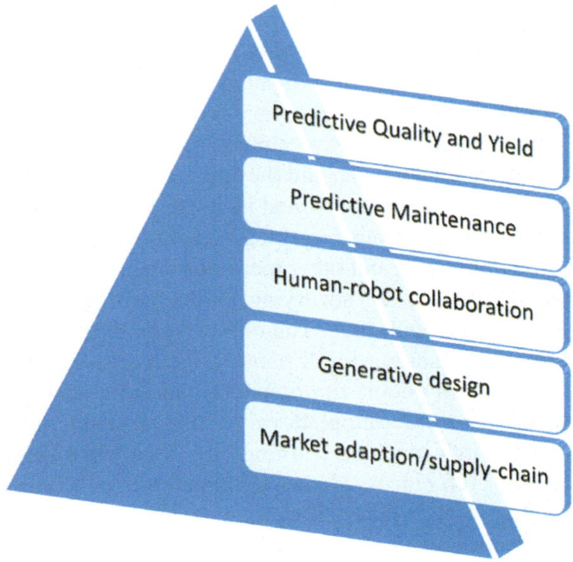

Fig. 2 Key drivers of artificial intelligence in food industry

2. Predictive Maintenance—One of the most basic and well-known Industrial AI applications is predictive maintenance. Predictive maintenance uses algorithms to anticipate the next malfunction of a component/machine/system instead of conducting maintenance according to a fixed timetable, and then warns workers to conduct focused maintenance procedures to avoid the failure, but not too early to waste downtime. The advent of Industry 4.0 and smart systems is contributing to increased exposure to solutions for predictive maintenance (PdM) that can minimize downtime costs and improve the availability (use rate) of production equipment [3].
3. Human—Robot Collaboration—The general strategy is that staff will be given training for higher-level roles in programming, architecture, and maintenance as jobs are taken over by robotics. In the meanwhile, as development robots are certified for work alongside humans, the quality of human–robot joint work is being increased. If the use of robotics grows in manufacturing, AI can play a significant role in ensuring the safety of human workers and allowing robots more responsibility for making decisions. The review deals with the implementation of cooperation between humans and robots, presenting the associated standards and modes of operation. There is a detailed literature review of works published in this field, with special attention to the key industrial applications [2].
4. Generative Designs—In the design level, designers will also make use of artificial intelligence. With a specifically defined design brief as feedback, designers and engineers may explore all the potential configurations of a solution using an AI algorithm, commonly referred to as generative design program. Restrictions and descriptions for content types, manufacturing processes, time limits and budget

restrictions should be included in the brief. You can then evaluate the set of solutions provided by the algorithm using machine learning.

5. Market adaptation—Artificial intelligence permeates the whole world of Business 4.0 and is not only confined to the manufacturing floor. The use of AI algorithms to improve the supply chain of manufacturing activities and to help them react faster to and predict business trends is one example of this. An algorithm can take into account demand patterns categorized by date, location, socioeconomic characteristics, macroeconomic activity, political status, weather pa, to generate estimates of market demand [4].

4 Applications of AI in Food Industry

Artificial Intelligence (AI) its advancement not only in food industry but also catching attention from all domains of business like technology to corporate houses, AI is controlling everywhere. Nowadays we cannot even imagine industries running without having good or minimum knowledge of AI. The food processing and handling industry (FP&H) will continue to grow at a very speedily rate till 2021 after that too and the reason is AI. The actual industry (FP&H) is running at a whopping above $100 billion and will continue to expand in coming decades.

The job opportunities creating due to AI and especially in the food processing and handling industry are at high rate as to maintain the technology even you need a high skilled and low skill people and packaging and handling almost require human intervention with great contribution of AI and technology to it making the complex things into very simpler for which human require lots and lots of efforts and time, but with the usage of AI and technology to the (FP&H) is making the industry growing at an unimaginable rates and will continue to do so in the future as well.

According to one report published by McKinsey [20], which says that the (FP&H) sector will grow at a constant rate with some ups and downs around 4–5% at least till 2021.

Classifying the overall benefits and importance of AI to (FP&H) Sector is divided into 5 broader segments, through which the complete cycle from starting to end rotates in the industry.

1. Package, sorting and handling
2. Hygiene and food safety measures as complied
3. Overall cleanliness and hygiene maintenance
4. Product development and its Management
5. Customer support and better decision making skills.

The subcategories and sub-division may be there but we can say and classify the overall management of (FP&H) sector worldwide via AI will come under this as from product handling to customer satisfaction and dealing it is included in all segments. All we can say is AI easing the human efforts but as advancement of AI these industries will need more people who have high skill set with understanding in machine, learning and technology due to which low skill worker are at a risk of losing their jobs.

We cannot say with this strongly that introduction of more AI into industries will take away jobs but somehow this will bring impact on the employment levels. The future will give more clear picture about this weather it's having a positive or negative impact.

4.1 Application of AI in Food and Packaging Industry (FP&H) Sector

No doubt food packaging and handling is a complicated process which require attention at each and every step to avoid any external contamination get involve with it. From raw material extraction and sorting, from hygienically storing them and using them for production of food product to customer satisfaction and finally delivery of the product.

Several processes are involved and many are interlinked with AI and technology these things are done at a very easily. Human intervention with technology makes the product more cheaply, better quality and with less time making the industry to produce more and earn more profits.

However, there are many FP&H industries where the complete process is driven by AI and technology and human intervention is there. Below are the 5 important applications of AI that are benefiting the food industry with increase in their revenue and better customer experience and services to provide.

1. **Package, Sorting and Handling**

This is very first challenge faced by the food industries is to sort the feedstock and operationally it is very time consuming. Accorded to several reports and food industry insights sorting are one of the tasks which involve majority of the work to be done by humans and involve their direct and indirect interventions to it.

For example: every farm based products or grown like potato, apple, orange and berries are different in shape and sizes and require rigorous sorting and thus it is also important to maintain to some important food standards and the quality check and the standards maintain by the food company.

Sorting is basically done by human labour and requires intense and continuous efforts to do so and these type of sorting was done till twentieth century and still in twenty-first century many companies and food industries are not enabled with AI, automation and technology. With the ease of AI, automation and technology the efforts and human intervention reduces to almost 0. Each type of sorting shape, size is done through machines more effectively and taking less time consumption which in short is better for the industry to produce more units.

TOMRA [25] Systems is one of the leading sorting and arranging providing solutions in the world based in Norway. TOMRA sorting food and recycling of products for better productivity and sustainable solutions and profits [12, 16]. Till twentieth century end 90% of the food sorting like of veggies, fruits and other food material were done through human hands which required great effort and more labour to work or long hours on a single sorting work making the process more complicated and time consuming.

After that introduction of AI, automation, IOT, Technology made the new cycle of process. Veggies, fruits are now being sorted through the machines. Best qualities of fruits were chosen from the better ones using x-rays and other machines to identify that. A new aspect to analyse the best qualities of veggies and fruits and which automatically made the sorting process easier and less time consuming than earlier. AI and IOT bring out the most helpful and needed advancement in the sector of food and processing because everything cannot be relied upon human and labours. Machines needs to be there for proper arrangement of products at large scale with proper sorting of its shape, size and order and bring out the best deal from it.

Japan based food processing and sorting company, KEWPIE Corporation created an AI technology based machines through which the anomalies and the other harmful factor and substances from the fruits, veggies and other things coming from farm directly. Through this technology the process becomes easier and the quality of manufacturing of products increases due to best quality of products. With the making of this new AI and IOT based technology these organizations and corporations are making other food processing units in profits and with production of high and good quality yield.

2. **Hygiene and Food Safety Measures**

Food safety and its linked measures is the concern for every food processing unit, corporation and organization. Everything needs to be perfectly executed in terms of hygiene of the product making, how it is making, under what conditions it is being made and what right ingredients to choose for making it an eatable product. Many questions are there and this will be there because this can risk to the health of many if precautions are not taken and followed seriously. Even the smallest to smallest contamination needs to be detected for the best product and thus outcome will be best and fits to all.

Food safety measures which include many things and concerns for organizations to get it followed on a serious like wearing of hand gloves while touching any food product and till and end of its package, how employees are handling the products and food packages and its distribution. To ensure all these safety concerns and food measures AI enabled cameras are being installed by the many food processing units and corporations. To keep check weather employees are maintaining and keeping all safety measure as said and governed. Through AI enabled sensors and cameras the managers of the organizations can keep a check all time. There are many food corporations, industries running without having proper safety and hygiene measures but what they don't know they are actually playing with human health and bringing out the contamination in the food products with their negligence and indiscipline. Somehow small scale industries need to put efforts towards these segments towards hygiene and food safety [23].

The Shanghai health and municipal corporation takes the charge with in collaboration and installed 200 AI based sensors and cameras into the restaurant's and food processing unit. With time the expansion of the 200 AI based cameras goes to 2000 and 20,000 food processing units and organizations. These advancements in the FP&H sector enabled the indiscipline at the real time and whether the food safety measures and regulations are followed or not [24].

3. **Cleanliness and Hygiene Maintenance**

Maintaining cleanliness is a strong concern for the food processing units and organizations. Companies want their product to be clean as like ice for the satisfaction of their customers and their retention too. Nowadays consumers are aware of these things and they want real check evidence that what are the safety and cleanliness measures followed during the making of the product. With the high using of AI, IoT, automation and technology food is being processed and picked through machines. But the concern is what if the machines parts are rusted or contaminated? This could be a serious concern with the health of consumers and safety measures of food are being neglected in that case. Though all things are being automated but still all things cannot be automated, some processes involve human touch and their intervention during the making of the product.

According to a study around 35% of water and energy [10] is being used only and only in the cleanliness process of the product and during its making. This 35% constitutes the most amount of expenditure for the organizations and food processing units. The large scale industries can somehow manage to do this task efficiently but what about the small scale units and corporations. As the industries are moving towards the automation, AI and IOT too these processes are now enabling through that and with its continuous usage and up gradation with time to time it's easy now to maintain the cleanliness standards by the food processing units and corporations. The budget goes down to 15–20% with respect to previously which was accounting to be 35%. This time with the help of AI and automation the companies are reducing their cost towards cleanliness and hygiene maintenance. It is estimated that companies will save around $150 million towards the reduction in the water usage and energy consumption.

Traditionally when AI was not enabled, food processing units and organization used to have a hand based cleaning systems which involves lots and lots of efforts and labour, eventually leading to higher cost for the organizations and food processing units to maintain the cost of cleanliness and hygiene. Though after spending lots of resources it was not guaranteed that the safety measures will look like as it was decided and governed because some food particles and pieces would leave behind and that could be the concern issue for the organizations and relating to food safety and measures. With the introduction of AI and its related advancement technologies to (FP&H) Sector many things have changed and are changing with that. Processes turned out to be in a very different way. Automation and AI enabled devices are used to clean the products; smallest to smallest details are being taken and carried out with help of this, sensors and cameras are being used to identify the contamination in product and concerning with its safety measures and protocols governed by the organizations. Self-cleaning equipment's using fluorescence images helps to identify many things and present debris in the any part of the equipment of food particle and making it free from all the contamination and waste of it. Traditionally from 19th—twentieth century to 21st many things have changed, majority of the task are now done through automation, AI enabled and IOT. As we are moving towards industry 4.0, which actually means everything needs to automated, self-driven and totally work should be done through machines and humans efforts and work should have turned out rom this to R&D for more development of these of automation and self-driven machines.

4. **Product development and its Management**

The (FP&H) sector is itself very unique and every time it needs innovation to satisfy the customer with something new products, but the question is how the companies get to know what to develop? And what kind of product needs to be launched in the market. Every time there must be something new, something connecting to the consumer day to day needs then only a product will up stand in the market and between the competitors. Before launching of AI, automation and IOT the companies needs to have campaign advertisement to check the taste of the consumer and what they actually want.

Coca Cola used to serve around 3000 + drinks around the world with 500 brands all over the world. Before AI was enabled they used to have a various campaign and advertisement around USA and all over to check what to launch next in the market, what type of drink customer would prefer, but after introduction of AI, automation and IOT to the world technologies and within Coca Cola changes started to happen and every things becoming automated and enabled through AI and its algorithm's. These types of algorithms are most important in deciding what to launch next and what kind of flavour to launch in the market. This type of logarithmic system requires lots of efforts to interpret the data but after enabling of AI these interpretation becomes so easy that launching of products in the market.

The Coca Cola enabled the fountain machine creating method for this using AI enabled technologies and software's which used for creating for a new soft drink by customer themselves. They just have to mix up the -two combinations of flavours like berries and cherries and based upon these choices Coca Cola decides to launch which of the best drinks in the market. These types of processes are very complicated if we involving human interventions to it but after installing of fountain drink combination machine across USA with enabling of AI and its automation self-driven technology, everything becomes very easy and reliable and the company the get the data of whatever they need whether it is in terms of new addition floured drinks, customer data, customer age using that fountain machine and which age group is preferring which type of flavour, all these important and relevant question came into existence only and only when there is AI, IOT and Automation self-driven technologies used. With using up of fountain drink machine Coca Cola came to know that majority of the population has mixed cherries with the sprite and later on with this data they invented and launched Cherry Sprite into the market and that too was a huge success because it was already known and developed through and by customers only, what they did is just collecting the data with the help of AI algorithm's and launched it.

AI, IOT and automation has transformed the processes and carrying out style of products which has also benefitted the organization to speed up their profits and revenues. Changes are difficult with AI for small scale manufacturing units and food processing units where in 2020 too majority of the work is being done by the labour and under their supervision too. People are relying upon AI and automation, adapting to their technologies and style but still it will take a long way to achieve complete automation in industries.

5. **Customer Support and Decision Making Skills**

From benefitting to food processing unit, AI also enables customer support and letting them know with their decision making for better purchase and buying. AI in decision making of customer is really an advancement to give a new experience to the customer, a new way of shopping. Shopping with your decision and a way you want to shop. This is what AI teaches customer and big brands. AI has enabled the various options and features in its graphic to enable customer a new way of shopping experience.

For example: Kellogg's [22] has introduced the AI support customer support with the help of IBM software's and technologies which enables customer to mix ingredients of 2 different kinds and letting them know whether they will go with each other or not.

5 Challenges faced by the Food Industry

AI, IOT and automation diverse in the food and processing industry will lead to greater advancements and heights in the industry. But as it is always said everything has its two faces the pros and the cons. If there are amazing technological benefits of food and processing units to manufacturing hubs but there are many challenges too for this, which cannot be neglected if the companies and Food processing units needs to grow and have a greater heights of revenue. Changes needs to be there in the AI and in the and in its implementation to the automation to industries and food processing units. Companies have to decide what kind of AI model they have to build for their units and it take it a lot to completely hold upon these systems.

Some of the major challenges of AI to Food Industry:

A. Cost and its Management—Potential to potential organization and food processing unit face the time to time up-gradation of the tech systems and it really cost too much for organization to contribute and take out the extra expenses for that. Buying this type of tech usually cost much and if their repair arises it's an extra expense to organization.
B. Technological Issues—there are many issues arising with new technologies and its integration. These types of systems are having network of integrations which

cannot be resolved so easily and require lot of expenses to fix that issue if so arises in the system or part of the machine. Sometimes the part once made for the machines are not meant further because lots of efforts are required and with day to day technology are changing and some or another thing will take over one next day.

6 Conclusion

Artificial intelligence is still at its initials round of taking over the complete industrial work especially in the food and processing unit. The sector requires lot and lot of changes with everything from top to bottom with manpower and to technological advancement's requiring resources and cost to begin with the process of AI in the sector. Growing up of AI and its advancement not only ensure the complete growth of the sector but a strong will to adapt these changes will bring out the growth in the (FP&H) sector. Adapting to AI and automation will require lots of efforts especially for the small scale food processing units. Sometime the cost will come as a drawback to adapt to these changes and sometimes the technological and integrity issues. AI and its automation will surely bring out the positive outlook and the speedily growth to the food and the processing sector. In the coming years the AI, automation and the IOT will going to revolutionaries the Food industries and its linked sector. Decade after decade the automation will take place in the industries and the day will come when the industries will work with full automation and no human intervention will be there. Only robots and technology will work for humans and humans will work for better AI and R&D towards it. Artificial intelligence (AI) will help companies to increase their revenues and profits, to cut down the maintenance cost of the unit, productivity efficiency will increase and a lot too customer experience by getting to know their likes, dislikes and what all new things they want from the companies, which will surely increase the positive way towards more development of new and innovative products.

There are many mobile development solutions also available interlinked with AI and its usage and now the bell has rung to step forward into the AI and the automated world. The world of complete automation. To stay in competition and to beat the competitors time to time change needs to be there to stand firmly in the market with strong roots of AI and automation. Food processing and handling has a billion-dollar business around the world every year and with the increase in population the business will further grow more, it is expected that by 2050 the world's population would reach up by 9 billion and to be able to satisfy the needs of such huge number's companies needs to have a very strong futuristic plan to implement upon. Constant changes towards the automation are required to achieve the growth which is required by the food industry to boost up their sales and revenues. Around the globe there are many example of American and western food which generally does have the same mixing of flavours which is generally done not by Asian dishes and the south Indian dishes. In order to find this change and why it is happening, intelligent science and data analytics is very useful to determine what consumer are wanting within the same flavour range. New innovations are there in industry and with the more usage of AI and IOT the industry will transform its style and way of presenting The Food processing and the handling sector are the important aspect to any economy and playing a great role in it. AI, IOT and automation has transformed the processes and carrying out style of products which has also benefitted the organization to speed up their profits and revenues. Changes are difficult with AI for small scale manufacturing units and food processing units where in 2020 too majority of the work is being done by the labour and under their supervision too. It totally depends upon how industries would take up this challenge to adapt to the new technological advancements and adopting their methods to artificial intelligence and IOT based. Sooner and gradually AI will have a very strong future which we all can see and with time everything is set to launch with customer preferences and taste which food industry wants too. People are relying upon AI and automation, adapting to their technologies and style but still it will take a long way to achieve complete automation in industries. Automation, AI and IOT will surely take over the industry to much more heights in the coming decades and there is the great scope of employment in the sector. Automation will require human interaction and its interaction will lead the world.

Growing up of AI and its advancement not only ensure the complete growth of the sector but a strong will to adapt these changes will bring out the growth in the (FP&H) sector. Adapting to AI and automation will require lots of efforts especially for the small scale food processing units. Sometime the cost will come as a drawback to adapt to these changes and sometimes the technological and integrity issues. AI and its automation will surely bring out the positive outlook and the speedily growth to the food and the processing sector. In the coming years the AI, automation and the IOT will going to revolutionaries the Food industries and its linked sector. Decade after decade the automation will take place in the industries and the day will come when the industries will work with full automation and no human intervention will be there. Only robots and technology will work for humans and humans will work for better AI and R&D towards it. Artificial intelligence (AI) will help companies to increase their revenues and profits, to cut down the maintenance cost of the unit, productivity efficiency will increase and a lot too customer experience by getting to know their likes, dislikes and what all new things they want from the companies, which will surely increase the positive way towards more development of new and innovative products.

There are many mobile development solutions also available interlinked with AI and its usage and now the bell has rung to step forward into the AI and the automated world. The world of complete automation. To stay in competition and to beat the competitors time to time change needs to be there to stand firmly in the market with strong roots of AI and automation. Artificial intelligence is taking the entire sector but food industry is the sector which will continuously be benefitting from the artificial intelligence, IOT and technological changes. Artificial intelligence (AI) will help companies to increase their revenues and profits, to cut down the maintenance cost of the unit, productivity efficiency will increase and a lot too customer experience by getting to know their likes, dislikes and what all new things they want from the companies, which will surely increase the positive way towards more development of new and innovative products. Taking into an account artificial intelligence and all the data science algorithms will surely have a more positive outlook in the food industry as it expected so. An estimate is there that 2050 food industry needs to grow its capacity by 2 times to feed the people and make their choices as well as preference's count. The new technologies of the AI and its interlinked machinery like IOT based, data science, taste preference and all other factor which were not imaginable before of AI but now the tables have turned and all we need is AI and IOT based desired product. Seeing the manufacturing outlets and industries growing more with robots and are working on the instructions as we feed and say them to do. To stay in the competition and to beat the market and the competitors company needs to revive with their strategy towards implementing of AI to their work field or not. We have many examples in the history which don't change with the technology and today those companies are struggling for their existence, like Nokia the king of Asian and European market till 90's but what mistake they made is not to adopt the new changes and go working with old patterns and strategies. Constant changes are the calling card for twenty-first century and company needs to constantly work upon these thing so that new emerging food industry, the drinks industry could shine out and have those kind of technologies integrated with AI to give better experience of food to customers.

7 Future of AI in Food Industry

Artificial Intelligence is no longer solely for the tech field. In several various market industries, innovative AI technology has been more commonly used, and the food service industry is capitalizing almost as well on computers as anything else. Your company will incorporate customer-facing technologies, integrate voice ordering, handle and monitor delivery orders with the most modern, up-to-date restaurant AI, and free up the employees to concentrate on cultivating the best possible customer experience.

Technology is progressing, and as it does, artificial intelligence can become a more common means of supplementing market practices and enhancing both the consumer engagement and your bottom line. The two are not mutually exclusive, much as AI technology does for your team, but rather work together. Customer retention and experience are two critically important facets of any good restaurant, and robots will not take over any of these. Machines will, however, work to grasp the nuances of market dynamics and other analytics, such that your workers focus their efforts in the most impactful fashion. AI and the human workers will boost profits together and make your organization the best it can possibly be.

8 Recommendations for Accelerating AI in Food Industry

AI is increasingly the latest frontier of digital innovation which, when integrated into workflows, is ideally placed to generate efficiency and effect across all facets of an enterprise. We agree that companies invested in the development of custom AI, as well as industry-specific solutions, are best placed to be tomorrow's global economic leaders.

1. In order to train and implement machine learning models cost-effectively and with the highest efficiency and scalability, our world class AI infrastructure.
2. Our AI building blocks, where we continue to invest in further strengthening our core skills in fundamental ways to understand images, speech, text, expression, and video.
3. It should be made easier for developers and data scientists to mark and prepare details, build models with Auto ML, tuning them and then manage their lifecycle with our AI Platform.

References

1. J. Schmitt, J. Bönig, T. Borggräfe, G. Beitinger, J. Deuse, Predictive model-based quality inspection using machine learning and edge cloud computing. Adv. Eng. Inf., **45**, 101101 (2020), ISSN14740346, https://doi.org/10.1016/j.aei.2020.101101
2. E. Matheson, R. Minto, E.G.G. Zampieri, M. Faccio, G. Rosati, Human–robot collaboration in manufacturing applications: a review. Robotics **8**(4), 100. MDPI AG. Retrieved from http://dx.doi.org/https://doi.org/10.3390/robotics8040100
3. W.J. Lee, H. Wu, H. Yun, H. Kim, M.B.G. Jun, J.W. Sutherland, Predictive maintenance of machine tool systems using artificial intelligence techniques applied to machine condition data. Procedia CIRP, **80**, 506–511 2019, ISSN 2212–8271, https://doi.org/10.1016/j.procir.2018.12.019
4. T.-C. Kuo, S. Smith, A systematic review of technologies involving eco-innovation for enterprises moving towards sustainability. J. Clean. Prod. **192**, 207–220 (2018)
5. V. Lakshmi, J. Corbett, How artificial intelligence improves agricultural productivity and sustainability: a global thematic analysis. In Proceedings of the 53rd Hawaii International Conference on System Sciences (HICSS Conference Office, Wailea, HI, USA, 2020)
6. B. Grieve, T. Duckett, M. Collison, L.A. Boyd, J. West, H. Yin, F. Arvin, S. Pearson, The challenges posed by global broadacre crops in delivering smart agri-robotic solutions: a fundamental rethink is required. Glob. Food Secur. **23**, 116–124 (2019)
7. Q. Sun, M. Zhang, A.S. Mujumdar, Recent developments of artificial intelligence in drying of fresh food: a review. Crit. Rev. Food Sci. Nutr. **59**, 2258–2275 (2018)
8. B.M. Boshkoska, S. Liu, G. Zhao, A. Fernandez, S. Gamboa, M. Del Pino, P. Zarate, J. Hernandez, H. Chen, A decision support system for evaluation of the knowledge sharing crossing boundaries in agri-food value chains. Comput. Ind. **110**, 64–80 (2019)
9. H. Barth, P.-O. Ulvenblad, P. Ulvenblad, Towards a conceptual framework of sustainable business model innovation in the agri-food sector: a systematic literature review. Sustainability **9**, 1620 (2017)

10. E.C. Ramírez, J.C. Albarrán, L.A.C. Salazar, The Control of Water Distribution Systems as a Holonic System, in *Service Oriented, Holonic and Multi-Agent Manufacturing Systems for Industry of the Future*. ed. by T. Borangiu, D. Trentesaux, P. Leitão, A. Giret Boggino, V. Botti (Springer International Publishing, Cham, Switzerland, 2019), pp. 352–365
11. Z. Hu, J. Pan, T. Fan, R. Yang, D. Manocha, Safe navigation with human instructions in complex scenes. IEEE Robot. Autom. Lett. **4**, 753–760 (2019)
12. E. Kosior, J. Mitchell, K. Davies, M. Kay, R. Ahmad, E. Billiet, J. Silver, Plastic packaging recycling using intelligent separation technologies for materials. In Proceedings of the Annual Technical Conference, Hyatt Regency, Santa Clara, CA, USA, (2017), pp. 500–506
13. R.J. Lehmann, R. Reiche, G. Schiefer, Future internet and the agri-food sector: state-of-the-art in literature and research. Comput. Electron. Agric. **89**, 158–174 (2012)
14. R.S. Mor, S. Singh, A. Bhardwaj, L.P. Singh, Technological implications of supply chain practices in agri-food sector-a review. Int. J. Supply Oper. Manag. **2**, 720 (2015)
15. N. Singh, P. Javadekar, Supply chain management of perishable food products: a strategy to achieve competitive advantage through knowledge management. Indian J. Mark. **41**, 10 (2011)
16. A. Renda, The age of foodtech: optimizing the agri-food chain with digital technologies. In *Achieving the Sustainable Development Goals through Sustainable Food Systems* (Springer Science and Business Media, Berlin/Heidelberg, Germany, 2019), pp. 171–187
17. https://www.seebo.com/industrial-ai/
18. L. Zhu, P. Spachos, E. Pensini, K.N. Plataniotis, Deep learning and machine vision for food processing: a survey. Curr. Res. Food Sci., **4**, 233–249 (2021), ISSN 2665–9271. https://doi.org/10.1016/j.crfs.2021.03.009
19. D. Saha, A. Manickavasagan, Machine learning techniques for analysis of hyperspectral images to determine quality of food products: a review. Curr. Res. Food Sci., **4**, 28–44 (2021), ISSN 2665–9271, https://doi.org/10.1016/j.crfs.2021.01.002
20. https://ppo.ca/mckinsey-on-the-food-processing-and-handling-industry/
21. https://www.ndtv.com/india-news/food-ordering-apps-zomato-swiggy-are-down-across-india-2866037
22. https://online.em.kellogg.northwestern.edu/
23. https://www.fooddocs.com/post/what-is-food-hygiene
24. I. Kumar, J. Rawat, N. Mohd, S. Husain, Opportunities of artificial intelligence and machine learning in the food industry. J. Food Qual., **2021**, ID 4535567, 10 (2021). https://doi.org/10.1155/2021/4535567
25. https://www.tomra.com/en/solutions/food

Transfer Learning-Based Models for Arabica Coffee Leaf Disease Classification

Mona M. Soliman and Aboul Ella Hassanein

Abstract Food quality and safety is a major concern for society since it is the foundation of human health, social progress, and stability. The growing concern about food safety in recent decades has increased the need for food quality monitoring worldwide. Nowadays, advances in artificial intelligence may substantially aid researchers and enterprises in boosting food processing efficiency. This paper concerns the coffee sector as one of the valuable commodities in the global economy. The coffee sector is subject to high levels of production volatility, caused not only by price concerns but also by a range of other uncontrollable factors, including inadequate crop management, weather events, and coffee pests and illnesses. One of the diseases that affect Coffee leaves is a miner disease. It has an impact on its production and quality. This study attempts to use different deep learning architecture approaches to spot this illness in coffee leaves and develop an expert system to help coffee farmers diagnose problems early on. Three deep learning models (e.g., ResNet18, GoogleNet, DenseNet161) are proposed to classify healthy and miner diseases in the coffee leaves. The best performances show an accuracy of 100% precision, recall, and F1 score for the GoogleNet model.

Keywords Deep learning · Transfer learning · Food quality · Coffee diseases · Arabica coffee

M. M. Soliman (✉) · A. E. Hassanein
Faculty of Computers and Artificial Intelligence, Cairo University, Cairo, Egypt
e-mail: mona.solyman@fci-cu.edu.eg

A. E. Hassanein
e-mail: Aboitcairo@cu.edu.eg

Scientific Research Group (SRGE), Cairo, Egypt

A. E. Hassanien and M. Soliman (eds.), *Artificial Intelligence: A Real Opportunity in the Food Industry*, Studies in Computational Intelligence 1000,
https://doi.org/10.1007/978-3-031-13702-0_7

1 Introduction

Food quality and safety is a complicated process that requires consideration of all phases of food processing, from cultivation, harvesting, and storage through preparation and consumption. These procedures, however, are frequently labor-intensive. This study aims to control food quality during the cultivation stage with a big concern given to the agriculture sector. Agriculture has always been a vital industry because of its economic influence, especially in developing nations. One of the most serious issues today is meeting the ever-increasing demands for high-quality food items. The world's population is rapidly growing, and with it comes an increased need for food and work. Farmers' conventional practices were unable to meet these needs. As a result, new automated approaches were developed. These new approaches fulfilled food demands while providing work possibilities for billions of people. Agriculture has transformed as a result of artificial intelligence. Artificial Intelligence (AI) has begun to play a significant part in our daily lives, expanding our senses and allowing us to alter our surroundings. Artificial intelligence-based techniques have safeguarded agricultural yields from various conditions, including climate change, population increase, job troubles, and food security concerns.

The main idea behind AI is to create technology that works similarly to a human brain. This technology is produced by researching how the human brain thinks, learns, makes judgments, and collaborates when addressing a problem, then developing intelligent software and systems on this foundation. These intelligent gadgets, like the human brain, are supplied with training data and provide us with the required result for every valid input. AI encompasses many fields, including Machine Learning and Deep Learning [1]. While AI is the science of developing intelligent systems, ML is the capacity to learn without being explicitly taught, and DL is deep neural network learning.

Convolutional neural networks (CNNs) are machine learning models that add depth and limitations to the early layers of a typical artificial neural network. The artificial neural network system supports DL technology (ANNs). Such ANNs are constantly learning algorithms, and by increasing the details, it is possible to improve the performance of the training processes. Larger data quantities are required for consistency. The training and inferring phases are the two main steps that make up the deep learning process.

In the field of agriculture, artificial intelligence (AI) is a new technology. AI-based technologies have elevated today's agricultural system to new heights. Crop productivity has improved, and real-time monitoring, harvesting, processing, and marketing have all been improved because of this technology [2]. The latest automated technology, such as agricultural robots and drones, have significantly contributed to the agricultural industry. Various high-tech computer-based methods are being developed to identify essential characteristics such as weed identification, yield detection, crop disease control, and many more techniques [3].

Even though various variables such as crop diseases, climate change, and many others directly impact food production, crop disease has been identified as one of the critical causes of food losses that disproportionately affect small-scale farmers [4]. According to [5], small-scale farmers generate roughly 80% of agriculture in underdeveloped nations, and Due to a scarcity of resources, food losses are higher in these places. More than 200 diseases (viruses, germs, and chemicals) are caused by tainted food, according to the World Health Organization [3], affecting 600 million people each year, with 420,000 deaths. Furthermore, emerging countries are finding it increasingly difficult to regulate and cure illnesses on time so that productivity and quality may be enhanced.

Coffee production worldwide is threatened by a range of coffee pests and diseases. The commercial coffee industry is centered on Robusta and Arabica coffee, with the latter accounting for over 70% of global production. The coffee industry faces significant output volatility, driven not only by price risks but also by various other uncontrolled hazards such as poor crop management, weather events (drought, unpredictable rainfall), and coffee pests and diseases.

This chapter suggests a deep learning algorithm for on-field disease identification and monitoring of Arabica coffee leaves to preserve plant health by preventing further infection. Using cell phones in a real-world setting will assist farmers in detecting early illnesses in arabica coffee leaves. The demand for vast volumes of data is driven by the fact that CNN training is an incredibly difficult optimization problem. To determine the minima for a loss function, they often employ stochastic gradient descent (SGD), which leverages huge labeled training datasets to minimise. Because SGD is a greedy strategy, finding the global minima is not guaranteed. This indicates that initialization can influence the ultimate result. These weights are typically obtained by sampling from a Gaussian distribution [6]. Compared to random initialization, parameter fine-tuning, a transfer learning strategy, has been shown to improve the performance of a CNN [7] greatly. The basic concept of transfer learning uses information from a related domain to another to improve an erudite. Deep le2arning and transfer learning, together with Graphics Processing Units (GPU) development, have resulted in a powerful categorization tool.

The proposed model utilizes three deep learning network architectures (e.g., ResNet18, GoogleNet, DenseNet161) to classify healthy and unhealthy coffee leaves. All these networks are based on using transfer learning. A collection of normal and aberrant photographs is collected in real-world settings. In Arabica coffee leaves, disease detection is a fine-grained classification issue in which a healthy leaf is separated from an unhealthy one based on small color and texture variations. JMuBEN2 [8] is a dataset of Arabica coffee leaves. The dataset contains two classes: healthy and unhealthy classes with miner disease. The miner disease is one of the most common diseases that afflict arabica coffee leaves. The image was taken on the Mutira coffee farm in Kenya's Kirinyaga district.

This chapter is organized as follows. Section 2 discusses the available relevant literature. Section 3 introduces the fundamentals of deep learning. Section 4 explains the suggested technique, findings, and discussion. Finally, Sect. 5 concludes the article and discusses future work.

2 Related Work

Deep learning is a relatively new, cutting-edge technology for image processing and data analysis with a lot of promise and potential. In recent years, deep learning has seen a lot of usage in agricultural applications for things like identification, classification, detection, quantification, and prediction. Convolutional neural network-based supervised learning is the most popular framework among deep learning models used in precision agriculture, as it can quickly and accurately extract the most effective characterization information from various features (color, shape, texture, size, spectral, venation, etc.) at different levels without requiring human expertise. Many purpose-specific datasets have been produced and provided freely, thanks to contributions from the agricultural research community. Early detection and immediate management of plant diseases and pests are critical for decreasing yield loss. In recent years, CNN-supervised learning has been extensively studied to develop automated, low-cost, accurate plant disease and pest classification and quantification algorithms [9–11]. Much DL-based agricultural research has benefited from PlantVillage [12], a publically available dataset released in 2015. It has 54,309 photos of healthy and sick plant leaves. A leaf from each plant was selected and placed on a paper sheet with a clean backdrop. A digital camera (Sony DSC-Rx100/13, 20.2 megapixels, automated mode) was used to take four to seven photographs at a standard location in natural light. Mohanty, et al. in [13] used two popular convolution networks, AlexNet, and GoogLeNet, to detect different crop species (up to 14) and about 26 illnesses in PlantVillage. Using an upgraded version of PlantVillage, [14] used 87,848 images in laboratory and farm regions to train and analyze five basic CNN architectures for identifying plant diseases [15]. The VGGNet achieved the highest success rate of 99.53 percent among the many CNN designs tested. However, because of this dataset's tiny number of plant and disease classes, it was still a long way from being a generic method that could be applied in real-world situations. [16] The performance of alternative learning techniques based on two pretraining tasks and three typical network architectures (VGG16, InceptionV3, and GoogLeNet) was compared, and it was discovered that the VGG16 model generalized better to unseen data. They discovered a peculiar phenomenon: networks trained using the crop–disease combination may focus on crop-specific features rather than disease regions, restricting the model's usefulness. As a result, they offered a more intuitive technique that concentrated just on illnesses, stating that it was more successful, especially when differentiating diseases from crops not included in the training set.

Authors in [17, 18] created a new dataset for classifying important multi-crop plant diseases. This dataset consisted of 121,955 photos collected using a mobile phone in real-world field circumstances during four growing seasons, and it comprised disease stages of 17 fungal infections on five crops that were nearly uniformly distributed (initial stages, early stages, medium stages, and advanced stages each account for a quarter of the total). After being segmented by a fully convolutional DenseNet network, each image was cropped to the region surrounding the principal leaf. The camera was set to manual focus, and the flash was turned off for all images.

Authors in [19] The disease severity of powdery mildew on cucumber leaves was assessed using a Remote Sens. 2022, 14, 559 6 of 27 DL-based pixel-wise semantic segmentation model. It's worth mentioning that just 50 pictures from the laboratory phenotyping platform were used in the study. As a result, despite the model's 96.08 percent average pixel accuracy, its performance in the real world must be verified. Garg K. et al. [20] proposed an end-to-end deep learning system called Cascaded MRCNN [20] for the automated measurement of Northern Leaf Blight (NLB) disease in field maize. They created the Severity Index (SI) to quantify the disease severity ratio. Seven thousand six hundred sixty-nine field maize photos obtained with a camera mounted on a DJI Matrice 600 UAV during the summer of 2017 were included in the collection. From the nadir perspective, the UAV was flown at the height of 6 m and a forward speed of 1 m per second, with images taken every 2 s. The sick areas were manually labeled using a VGG Image Annotator. The experimental findings indicated a 73 percent correlation between the predicted and ground truth SI on a field maize dataset with NLB illness.

3 Deep Learning

Deep Learning (DL)enhances conventional ML by adding more "depth" (complexity) to the model and modifying the data using various functions that hierarchically allow data representation through many layers of abstraction [21]. Feature learning, or the automated extraction of features from raw data, is a key benefit of DL, with features from higher levels of the hierarchy produced by the composition of lower-level features [22]. Because of the more complicated models utilized in DL, which allow huge parallelization, it can tackle more difficult problems exceptionally well and quickly. If there are sufficiently substantial datasets characterizing the problem, these complicated models used in DL can improve classification accuracy or minimize error in regression situations.

Convolutional Neural Networks (CNN) are deep, feed-forward artificial neural networks. Different convolutions are applied to different levels of the network, resulting in diverse representations of the learning dataset, starting with more generic ones at the initial bigger layers and progressing to more specialized ones at the deeper layers. The convolutional layers extract features from the input pictures, which are decreased in dimensionality by the pooling layers. The convolutional layers encode multiple lower-level characteristics into more discriminative features in a spatially context-aware manner [23].

Input, hidden, and output layers are the main components in CNN models. The input layer allows the network to receive data from the outside world. The hidden layers perform calculations and data transmission from the input to output layers. The output layers take input from the CNN network, a learned representation of the input pictures, and categorize them [24]. In the network, information is fed forward (from the input, through the hidden layers, to the output). The loss is determined using the cross-entropy function once the forward pass is done, and the model weights are adjusted to improve the forecast for the following pass. The loss is determined using the cross-entropy function when the forward pass is completed, and the model weights are adjusted to improve the forecast for the following pass. The least loss point is found by performing this forward pass across several epochs. The model may learn the link between the feature and the goal by minimizing the error between the model prediction and the intended output [25]. The model may learn the link between the feature and the target by minimizing the error between the model prediction. The intended output is given a collection of features in x and the desired output y.

Convolution, pooling, and fully connected layers are the three essential components of CNN architecture. Convolution is performed to the input data (x) using a convolution filter/kernel (w) in the first layer to build a feature map (s). A convolution layer is often made up of numerous kernels that conduct several convolutions on the input. Each convolution operation employs a separate kernel, yielding a unique feature map. The collected feature maps are piled together to form the final output of the convolution layer. When more convolutional layers are added, the model may learn characteristics hierarchically (gradients, lines, edges, basic shapes, complex shapes, etc.)

Another typical process is a pooling operation, which maps a subregion to its maximum value (maximum pooling) or average value (average pooling) (average pooling). The sensitivity of the convolution result to image variation is reduced using this method (e.g., in object translations). Another basic procedure is using an activation function, such as the rectified linear unit (ReLU). ReLU adds nonlinearity to deep learning models. It takes an integer as input and converts negative numbers to zeros. In mathematics, ReLU is expressed as y 14 max (0, x). The photo classification problem will be handled linearly if this function is not utilized. This layer helps reduce the number of parameters in the system [26].

Feed forward neural networks are what the Fully Connected Layer is all about. The output from the final Pooling or Convolutional Layer, which is flattened and fed into the fully connected layer, is the input to the fully connected layer. A matrix is converted into a column vector using the flattened layer. To improve the network's computational efficiency and stability throughout training. After passing through the fully connected layers, the final layer employs the softmax activation function (rather than ReLU) to determine the likelihood that the input belongs to a specific class (classification). Finally, we get the probability that the item in the picture belongs to each category [27].

4 Experimental Design, Materials, and Methods

4.1 Data Material

Coffee is a member of the Rubiaceae family and is produced in vast plantations where high-value fruit species are prioritized. From a 25% to 100% yield. Sun-grown coffee will out-yield shade-grown coffee if members of the genus Coffea, two key species, are fertilized more widely [28].

Coffee trees and shrubs are small evergreen trees or shrubs with smooth leaves and many stems. The leaves are dark, glossy green, and oval. Coffee comes from Africa and is known as Arabian coffee (*C. arabica*) or robusta coffee (*C. cenephora*). Coffee plants produce creamy-white flower clusters and a berry-like fruit with two seeds, which is generally referred to as a berry. The fruit starts off green, but as it ripens, it turns scarlet red and then black when dried. *C. canephora* grows to a height of 15 m (49 ft), but *C. arabica* grows to just 4–5 m (13–16 ft). The trees have a 20–30 year lifespan.

Where possible, Coffea arabica self-pollinates despite only catching 3–13 percent of incident rainfall. It grows best in sandy to loamy soils with plenty of sunshine. The correlations between climatic parameters are generally textural, with pH 5–6 being mildly acidic [29]. Because many factors influence coffee production and quality, such as soil water status, environmental factors influence growth and development of climatic conditions, agricultural management systems for plants under different forms during the growth stages of the coffee crop, agricultural management systems for plants under different forms during the growth stages of the coffee crop, agricultural management systems for plants under different forms during the growth stages of the coffee crop, agricultural management systems for plants under different forms during the growth stages of the coffee crop, agricultural management systems for plants under different forms during the growth stages of the coffee crop, agricultural One of the cultivars' characteristics is coffee shading [30, 31].

JMuBEN2 [8] Arabica coffee leaf datasets are described in this article. Using a digital camera and the assistance of a pathologist, images were captured on the Mutira coffee farm in Kirinyaga county, Kenya, in real-world conditions. JMuBEN2 is split into two zipped files, one with 16,979 photographs of Miner and the other with 18,985 images with healthy leaves. According to the pathologist, yellow tracks are left beneath the coffee leaf epidermal while miner larvae eat. With the description in hand, the plant pathologist walked across the coffee farm, identifying the damaged leaves by observation [32]. Figure 1 depicts photos of healthy and miner people. Table 1 shows the distribution of training and testing samples.

4.2 Data Pre-processing

To increase image quality, image misrepresentations and sounds were removed during the pre-processing phase. Noise filtering and contrast stretching, both done with mean and median filters, were used during data pre-processing.

Data augmentation was performed on field-collected photographs to improve smaller datasets by translating them into larger ones. Rotation and flipping were utilized as data augmentation techniques in this study. The rotation was done at 180° horizontally and vertically to accentuate the disease-affected region of concern. The image rotator tool was used to rotate the picture counterclockwise. An image flipper

Fig. 1 Dataset samples (Healthy versus Miner)

Table 1 Dataset distribution for training and testing

	Training	Testing
Healthy	8000	2000
Miner	8000	2000
Total	16,000	4000

was used in the instance of flipping. A horizontal flipper was used to flip certain pictures horizontally, while a vertical flipper was used to flip others vertically. The photos were flipped to provide for a better display of the sick areas. Figure 2 shows some examples of data augmentation process.

Fig. 2 Data augmentation examples

4.3 Model Description

To categorize healthy and miner coffee leaves, the proposed model uses three deep learning network architectures (e.g., ResNet18, GoogleNet, DenseNet161). Transfer learning is used in all of these networks. In real-life scenarios, a collection of normal and abnormal photos is gathered. The proposed model aims to classify healthy and miner images in JMuBEN2 [8] dataset of Arabica coffee leaf. Figure 3 shows a

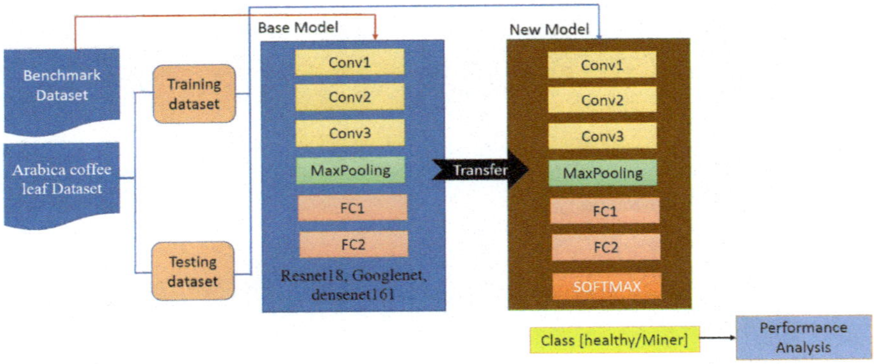

Fig. 3 Arabic coffee leaf classification model using transfer learning

generic classification model using the concept of transfer learning with three different networks (e.g., ResNet18, GoogleNet, DenseNet161).

CNNs are made from a combination: a convolution layer that splits the image's numerous elements for analysis and a fully connected layer that takes the output of the convolution layer and predicts the best classification for the image. Since 2012, many CCN architectures with varied architectures have been suggested [23]. Transfer learning enables an existing model's final layer to be retrained, resulting in a considerable reduction in training time and dataset size. Retraining the last layer allows you to save the information that the model gained during its initial training and apply it to a smaller dataset, resulting in highly accurate classifications without requiring substantial training or computer capacity. Resnet18, Googlenet, and densenet161 models are among the most well-known models employed for transfer learning in this study. The model takes coffee leaf images as input and classifies each as healthy or miner.

- **Resnet18 Transfer Learning Based Model**

We begin with a pertained residual neural network model (ResNet18), which can recognize thousands of different item categories. This model was previously trained on 1.5 million photos (ImageNet), which include various images. ResNet18's overall design is seen in Fig. 4. In its most basic form, a residual neural network model may be stated mathematically as:

$$y = f\left(\mathrm{h(x)} + \mathrm{g_b(x)}\right) \tag{1}$$

where (x, y) is the input–output data, p are the adjustable parameters of the function ***gp***, and f and h are fixed functions. The input data in our situation is an image described as a three-dimensional array that includes the picture's pixels and color channels [21]. Equation (1) describes an optimization procedure for identifying unknown parameters of functions. This research uses a stochastic gradient descent-based technique

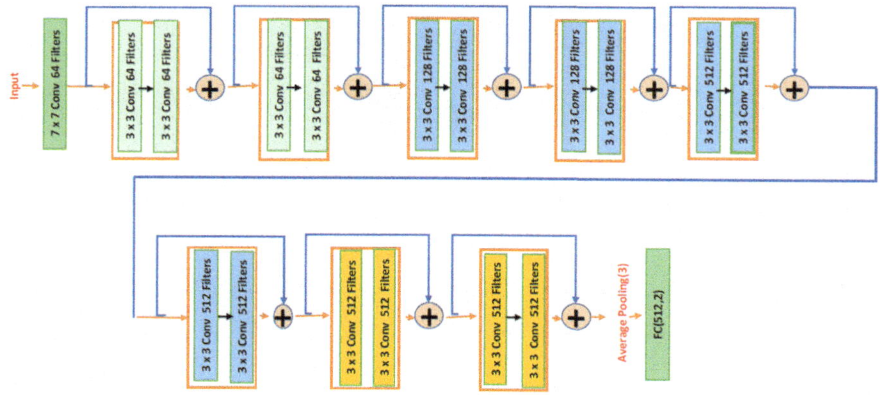

Fig. 4 Basic architecture of ReseNet18

(AdamW) with deployed backpropagation. To implement the model, we utilize the standard PyTorch and fastai libraries and the ICEXPERT source code files. Then, we train and test our model based on its accuracy using the Arabic coffee leaf dataset.

- **GoogleNet Transfer Learning Based Model**

GoogleNet, also known as Inception-V1, has the primary goal of achieving high accuracy while minimising processing costs. It presented a unique inception block strategy in CNN as shown in Fig. 5. Inception block t uses the divide, transform, and merge paradigm to combine multi-scale convolutional transformations [33]. This block has filters of various sizes (1, 3×3, and 5×5) for collecting spatial information at different scales (fine and coarse grain levels) [34]. In GoogleNet, convolutional layers are modified in small chunks, similar to the Network in Network (NIN) architecture's concept of replacing each layer with micro NN [34]. GoogleNet's use of the split, modify and merge idea contributed to resolving a problem requiring the recognition of several variants within the same category of different photos. In addition to enhancing learning capacity, GoogleNet enhances the CNN parameter's efficiency [34]. Nine inception modules, 22 layers, four max pooling layers, and one average pooling layer make up the basic GoogleNet design. The ReLU is used by all convolutional layers, including those inside the inception modules. During the training phase, two auxiliary classifiers are added to the layers in the network's core to deal with the problem of vanishing gradients. The softmax classifier assigns a dropout ratio of 0.4 [35].

- **DenseNet161 Transfer Learning Based Model**

The goal of Densenet was to solve the problem of vanishing gradients in high-level neural networks, which result in lower accuracy. Simply said, due to the lengthier route between the input and output layers, information evaporates before it reaches its destination. DenseNets provide some benefits, including improved information flow

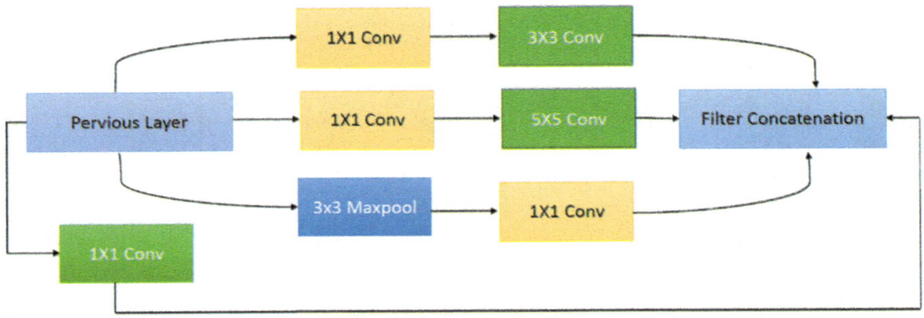

Fig. 5 Basic architecture of the inception block

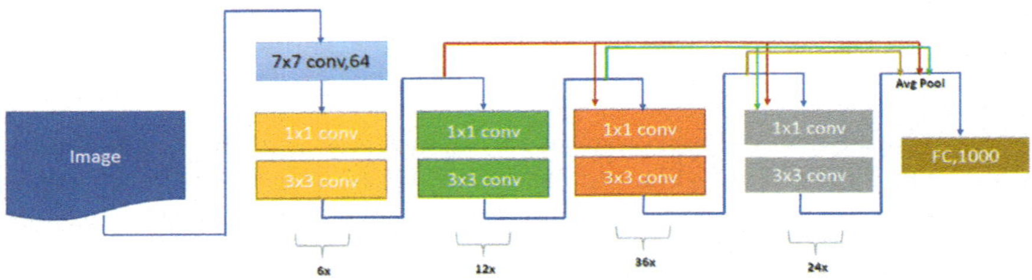

Fig. 6 Basic architecture of DenseNet-161

and gradients throughout the network, making them easy to train. The loss function and original input signal gradients are directly accessible to each layer, resulting in implicit deep supervision [36]. This makes it easier to train more complicated network topologies. Dense connections also appear to have a regularising effect, preventing overfitting tasks with smaller training sets. DenseNet integrates characteristics from all preceding convolutional blocks, whereas ResNet only uses one prior feature map. Both ResNet and DenseNet share the principle of linking to the feature maps of all previous convolutional blocks [37]. DenseNet takes longer to train because it uses many little convolutions in the network, which run slower on GPU than a big compact convolution with the same number of GFLOPS.

DenseNet-161 is the largest model in the DenseNet group, weighing over 100 MB. As shown in Fig. 6, the DenseNet-161 model is built up of four dense blocks, with all of the layers frozen before the third block. This ensures that the model can extract the image's crucial features while also lowering training time.

4.4 Results and Analysis

Several tests with the trained model were done with the test images to check the algorithm's performance. Precision, Recall, and F1 Score are the three main measures

Table 2 Confusion matrix

Labeled	Classification result	Confusion matrix
Positive	Positive	TP
Positive	Negative	FP
Negative	Positive	FN
Negative	Negative	TN

Table 3 Accuracy measures for coffee leaves classifications

		Precision	Recall	F1-score	Accuracy	Support
Resnet18	H	0.86	0.52	0.65	0.53	3338
	M	0.19	0.59	0.29		662
Googlenet	H	1	1	1	1	2000
	M	1	1	1		2000
Densenet161	H	1	0.79	0.88	0.86	2543
	M	0.73	1	0.84		1457

for measuring the performance of neural network models. True positive (TP), false positive (FP), true negative (TN), and false negative (FN). For binary classification problems, samples can be divided into four types: true positive (TP), false positive (FP), true negative (TN), and false negative (FN), according to the combinations of the true class and predicted class of the learner. Table 3 shows the confusion matrix for the categorization results. Table 2 Confusion matrix for the classification results.

Precision(P) and recall(R) are defined as follows:

$$Precision = \frac{TP}{TP + FP} \tag{2}$$

$$Recall = \frac{TP}{TP + FN} \tag{3}$$

The F1 score was also used to evaluate the performance of the model. The definition of the F1 score is shown as follows:

$$F1 = \frac{2 * precision * recall}{precision + recall} \tag{4}$$

Table 3 shows comparative classification measures for the three proposed networks: ResNet18, GoogleNet, and DenesNet161. The accuracy measures are evaluated for both healthy (H) and Miner (M) classes. The best accuracy measures are achieved within the GoogleNet network. It can classify healthily, and miner coffee leaves with accuracy measures reaching 100% for precession, recall, and F1-score respectively for 2000 samples used for training and 2000 samples for testing. Resnet18 gives 53% overall accuracy for 3338 training samples and 662 testing samples, while densenet161 gives 86% overall accuracy for 2543 training samples and 1457.

Figure 7 depicts the training accuracy/loss curves for the three suggested models. The training accuracy/loss curves for all models decline to a stabilization. The error on the training data for the models is decreased over epochs.

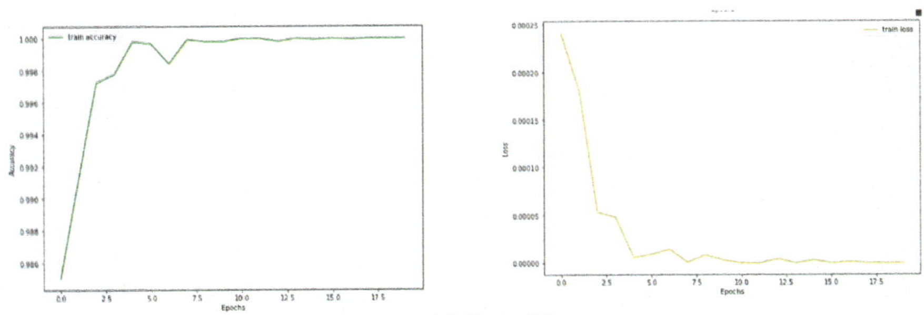

(a) Resnet18

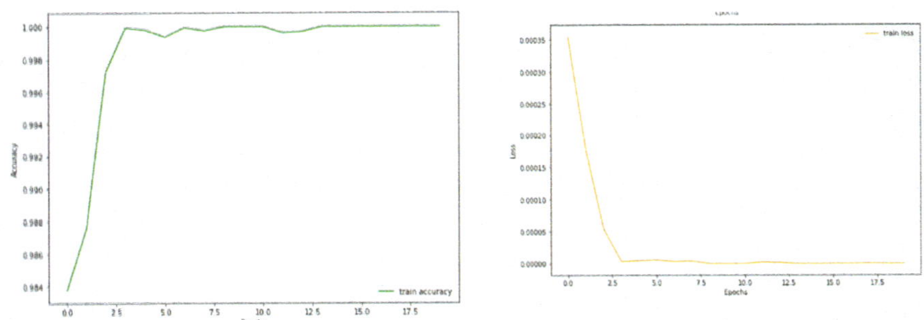

(b) Googlenet

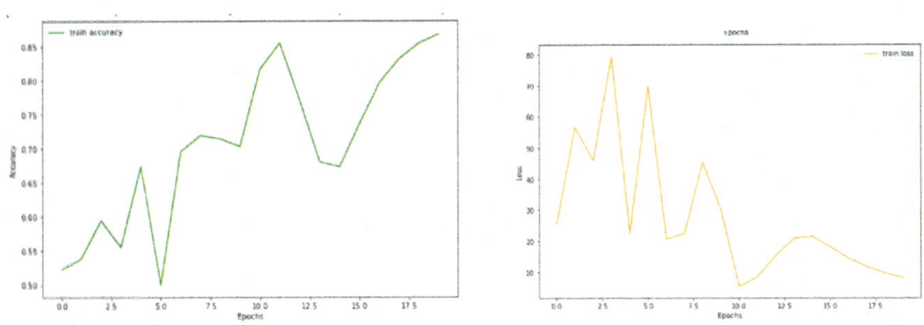

(c) densenet161

Fig. 7 Training/loss curves for deep learning network **a** Resnet. **b** Googlenet. **c** Densenet161

5 Conclusions and Future Work

This research looked at computer approaches for detecting illnesses in coffee leaf images to assist coffee farmers in increasing their productivity and improving the quality of their products. In this paper, we presented a transfer learning model based on Resnet18, Googlenet, and densenet161. This approach was used to classify arabica coffee leaves into healthy and ill classifications. The recommended model achieved the highest accuracy in the training phase using Googlenet and a small dataset. More samples, characteristics, and models were required for this study. More illnesses, rather than just two, will be categorized in the future to support coffee producers in boosting their productivity and product quality. This study might be expanded by developing a comprehensive expert system for rating illnesses in Arabic coffee leaves on different days.

References

1. D. Patel, D. Shah, M. Shah, The intertwine of brain and body: a quantitative analysis on how big data influences the system of sports. Ann. Data Sci., (2020). https://doi.org/10.1007/s40745-019-00239-y
2. H. Yang, W. Liusheng, X.H. Junmin, Wireless Sensor Networks for Intensive Irrigated Agriculture, Consumer Communications and Networking Conference, 2007. CCNC 2007, Las Vegas, Nevada, 4th IEEE (2007), pp. 197–201
3. K. Liakos, P. Busato, D. Moshou, S. Pearson, D. Bochtis, Machine learning in agriculture: a review. Sensors **18**(8), 26–74 (2018). https://doi.org/10.3390/s18082674
4. J. Ahmad, B. Jan, H. Farman, W. Ahmad, A disease detection in plum using convolutional neural network under true field conditions. Sensors **20**, 5569 (2020). https://doi.org/10.3390/s20195569
5. IFAD, *Smallholders, Food Security, and the Environment* (International Fund for Agricultural Development, Rome, Italy, 2013)
6. K. Philipp, D. Carl, J. Donahue, T. Darrell, Data dependent initializations of convolutional neural networks. In: International Conference on Computer Vision, arXiv:1511.06856v3, (2016), pp. 1–12
7. J. Yosinski, J. Clune, Y. Bengio, H. Lipson, How transferable are features in deep neural networks? In Advances in Neural Information Processing Systems (Proceedings NIPS), vol 27, (2014), pp. 1–9
8. J. Jennifer, M. David, K. Benson, C. Edna, JMuBEN2. Mendeley Data, V1, (2021). https://doi.org/10.17632/tgv3zb82nd.1
9. S.K. Noon, M. Amjad, M.A. Qureshi, A. Mannan, Use of deep learning techniques for identification of plant leaf stresses: a review. Sustain. Comput. Inform. Syst. **28**, 100443 (2020)
10. W.H. Zeng, M. Li, Crop leaf disease recognition based on self-attention convolutional neural network. Comput. Electron. Agric. **172**, 105341 (2020)
11. E.C. Too, L. Yujian, S. Njuki, L. Yingchun, A comparative study of fine-tuning deep learning models for plant disease identification. Comput. Electron. Agric. **161**, 272–279 (2019)
12. D.P. Hughes, M. Salathé, An open access repository of images on plant health to enable the development of mobile disease diagnostics., (2015) arXiv:1511.08060
13. P. Sharada, P.H. David, S. Marcel, Using deep learning for image-based plant disease detection . Frontiers Plant Sci., **7** (2016). https://doi.org/10.3389/fpls.2016.01419

14. P. Bedi, P. Gole: Plant disease detection using hybrid model based on convolutional autoencoder and convolutional neural network. Artific. Intell. Agricult., **5**, 90–101 (2021). ISSN 2589–7217, https://doi.org/10.1016/j.aiia.2021.05.002
15. K.P. Ferentinos, Deep learning models for plant disease detection and diagnosis. Comput. Electron. Agric. **145**, 311–318 (2018)
16. S.H. Lee, H. Goëau, P. Bonnet, A. Joly, New perspectives on plant disease characterization based on deep learning. Comput. Electron. Agric. **170**, 105220 (2020)
17. A. Picon, M. Seitz, A. Alvarez-Gila, P. Mohnke, A. Ortiz-Barredo, J. Echazarra, Crop conditional convolutional neural networks for massive multi-crop plant disease classification over cell phone acquired images taken on real field conditions. Comput. Electron. Agric. **167**, 105093 (2019)
18. A. Picon, M. Seitz, A. Alvarez-Gila, P. Mohnke, A. Ortiz-Barredo, J. Echazarra, Deep convolutional neural networks for mobile capture device-based crop disease classification in the wild. Comput. Electron. Agric. **161**, 280–290 (2019)
19. K. Lin, L. Gong, Y.X. Huang, C.L. Liu, J.H. Pan, Deep learning-based segmentation and quantification of cucumber powdery mildew using convolutional neural network. Front. Plant Sci. **10**, 155 (2019)
20. K. Garg, S. Bhugra, B. Lall, Automatic quantification of plant disease from field image data using deep learning. In Proceedings of the IEEE/CVF Winter Conference on Applications of Computer Vision, Waikoloa, HI, USA, (2021), pp. 1965–1972
21. Y. LeCun, Y. Bengio, G. Hinton, Deep learning. Nature **521**(7553), 436–444 (2015)
22. Y. LeCun, Y. Bengio, Convolutional networks for images, speech, and time series. The Handbook Brain Theory Neural Networks, **3361**(10) (1995)
23. M. Soliman, A. Darwish, A.E. Hassanien, Deep learning technology for tackling COVID-19 pandemic. In: Digital Transformation and Emerging Technologies for Fighting COVID-19 Pandemic: Innovative Approaches. Studies in Systems, Decision and Control, vol 322 (Springer, Cham, 2021). https://doi.org/10.1007/978-3-030-63307-3_9
24. I. Goodfellow, Y. Bengio, A. Courville, *Deep Learning* (MIT Press, 2016)
25. M.D. Zeiler, R. Fergus, Visualizing and understanding convolutional networks, in: *European Conference on Computer Vision*, (2014), pp. 818–833
26. E. Kim, G.S. Dahiya, S. Løset, R. Skjetne, Can a computer see what an ice expert sees? Multilabel ice objects classification with convolutional neural networks, Results in Engineering, vol 4, ISSN 2590–1230, (2019), https://doi.org/10.1016/j.rineng.2019.100036
27. Sh. Basha, D.S. Ram, P. Viswanath, M. Snehasis, Impact of fully connected layers on performance of convolutional neural networks for image classification. Neurocomputing. **378** (2019). https://doi.org/10.1016/j.neucom.2019.10.008
28. Z.Q. Cai, Y.J. Chen, Y.H. Guo, K.F. Cao, Responses of two field-grown coffee species to drought and rehydration. Photosynthetica **43**, 187–193 (2005)
29. F.M. DaMatta, C.P. Ronchi, M. Maestri, R.S. Barros, Ecophysiology of coffee growth and production: a review. Braz. J. Plant Physiol. **19**, 485–510 (2007)
30. R.N. Mayoli, K.M. Gitau, The effects of shade trees on physiology of Arabica Coffee. Afr. J. Hort. Sci. **6**, 35–42 (2012)
31. M.B.P. Camargo, The impact of climatic variability and climate change on Arabic coffee crop in Brazil. Bragantia **69**, 12 (2010)
32. M.D.C. Alves, L.G. De Carvalho, E.A. Pozza, L. Sanches, J.D.S. Maia, **Modelling plant** disease risk areas based on Brazilian climate change scenarios. J. Environ. Sci. Eng. **5**(3), 333–343 (2011)
33. [Basic-5] C. Szegedy, W. Liu, Y. Jia, P. Sermanet, S. Reed et al., Going deeper with convolutions. In Proceedings of the IEEE Conference on Computer Vision and Pattern Recognition, Jun. 7–12, IEEE Xplore Press, Boston, MA, USA (2015). https://doi.org/10.1109/CVPR.2015.7298594
34. GoogleNet CNN Neural Network towards Chest CTCoronavirus Medical Image Classification 1Nesreen Alsharman and 2 Ibrahim Jawarneh, Journal of Computer Science, **16**(5), 620–625 (2020)

35. P. Pornntiwa, O. Emmanuel, S. Olarik, S. Lambert, W. Marco, Comparing local descriptors and bags of visual words to deep convolutional neural networks for plant recognition (2017), https://doi.org/10.5220/0006196204790486
36. G. Huang, Z. Liu, L. Maaten, Densely Connected Convolutional Networks, Densely Connected Convolutional Networks, arXiv:1608.06993v5 (2018)
37. G. Huang, S. Liu, L. Maaten, K.Q. Weinberger, An efficient densenet using learned group convolutions, arXiv:1711.09224v2 [cs.CV] (2018)

Potential of IoT for Water Monitoring to Upgrade Food Quality

Medini Gupta, Sarvesh Tanwar, and Vishal Jain

Abstract The Internet of Things (IoT) is used for tracking, gathering, and analyzing data from remote locations in various research areas. Because of the massive rise in global industrialization production, small town to city migration, and overuse of state land and resources of the ocean, the standard of water accessible by humans has significantly degraded. Around 5 million people worldwide die from waterborne diseases (Water Management Information System of India, 2017). The bad quality of water spreads disease, causes loss of life, and hinders socioeconomic development. On the other hand, infrastructure such as pipes deteriorates due to water aging. Under such circumstances, modern automation in the infrastructure of the water was needed to provide the delivery of great standard water to consumers securely and at a low in cost way, to make efficient use of the precious water resources of our world. Water is an important requirement for human continuity. Thus, systems must be placed to test strenuously the water standard at hand for consumption in industrial, natural recommended sources, including the water bodies like shores surrounding our municipality and metropolis. Water is used in different food production levels, and poor water quality can also affect food and consumer health. Food quality can be monitored in real time by converging different IoT technologies. A provision that a high standard of water is important in the prevention and spread of waterborne disease and the enhancement of quality of life.

Keywords Internet of things · Global industrialization · Modern automation · Waterborne disease · Smart meters · Food quality

M. Gupta · S. Tanwar (✉)
Amity Institute of Information Technology, Amity University Uttar Pradesh, Noida, India
e-mail: s.tanwar1521@gmail.com

V. Jain
Department of Computer Science and Engineering, School of Engineering and Technology, Sharda University, Greater Noida, India

A. E. Hassanien and M. Soliman (eds.), *Artificial Intelligence: A Real Opportunity in the Food Industry*, Studies in Computational Intelligence 1000,
https://doi.org/10.1007/978-3-031-13702-0_8

1 Introduction

The water we can use on the planet is a precious asset and comprises just 0.01% of the earth's available water. Analyzing the standard of water we can intake is important before consumption in everyday life because it can directly or indirectly affect human health. As many as 80% of diseases are associated with low water quality and sanitation [1]. The water available is highly contaminated, and the situation could get worse in the future. The water demand supply balance is projected to become powerful from now on because of the growth of the population from all over the world. The water system is a big network comprised of many processes, including water consumption, purification, consumer delivery, sewage disposal, etc. [1]. Certain causes of water pollution include floods and droughts and lack of consumer knowledge and education. Water nearby Fiji had slowly failed to resist pressure to a good extent of poor pollution quality over the last few decades. Contaminated material and leakage from oil are some main, central water pollution modes that threaten Fiji's water sources. A report printed in Fiji's Times in 2014, for instance, stated that the flow of raw sewage into the Samabula river at a speed of 300 L in a second because of damaged pipes. Completely removing emissions may seem like an unbelievable idea, but it is certainly possible to minimize its consequences when it does happen [2].

The main purpose of this chapter is to set up a system for monitoring seawater quality to help control Fiji water pollution using IoT. Therefore, various methods are used to track the consistency of the water in real time. The Smart Water Analysing project will calculate the following analytical water framework; Oxidation and Reduction Potential (ORP), Potential Hydrogen(pH), Conductivity, and Temperature. When tracking certain parameters, it is considered that a constant set of results should be achieved [3, 4]. Therefore, a continuous sequence of anomalous measurements will signify the possible presence of water pollutants, and with the aid of IoT technology, the consumer would be informed of this operation. Short-term anomalous readings will be reported but not interpreted as a warning. Thus, with this monitoring strategy effectively implemented, an alert system for water contamination can be accomplished with a completely developed network using several monitoring stations. Smart Water Monitoring System is an essential human resource. The drastic climate change and population growth caused water to be inadequate compared with the growing demand. Water management becomes a major challenge for both government and water companies as water use increases.

Water is the most critical asset of life. One of the global problems is water pollution. Level of water is going down in the current scenario [5]. Due to advances in information technology, there is an increase in the importance of intelligent solutions for controlling water quality. Water quality should be monitored regularly to ensure that a safe water supply is consumed and water is used for multiple reasons, such as irrigation.

2 Literature Review

IoT has been widely used in the water industry. Due to the ever-growing population and scarce resources, it is very difficult for water service providers to deliver potable water. Lack of analytical tools and monitoring systems along with labor force prevents them from increasing their productivity and identifying problems to be resolved. Table 1 also elaborates on the background work carried out by different authors in past years.

Vennam Madhavireddy, Koteswarrao [9] proposed a system that collects the data about various parameters such as CO_2, temperature, pH, and water level. The device gets triggered when the micro-controller is switched ON and identifies the water parameters of respective sensors. Through the Wi-Fi module, the data from water parameters are transferred wirelessly on a web server and can be retrieved from anywhere. All the processes are taking place in real time. The data is refreshed every 5 s. For tracking the sensors, one hour is chosen. This reduces power usage. There is an increase in system efficiency and a decrease in complexity.

Sathish Pasika and Sai Teja Gandla [10] presented a system using IoT to monitor drinking water quality. It is an effective and affordable system to monitor real-time water quality. It comprises various sensors for measuring different parameters, including humidity and temperature in the environment, pH level, water tank level, and turbidity. The Microcontroller Unit is attached to these sensors. Personal Computer is used to monitor the obtained information. ThinkSpeak is an IoT-based program used to measure water quality, and the gathered information is delivered to the cloud. ThinkSpeak application can be installed over the internet. It requires the user's ID and password to access the stored information.

Monira Mukta et al. in [11] discussed the role of IoT in measuring four physical parameters such as electric conductivity, temperature, pH, and turbidity for monitoring smart water quality systems. All four sensors for measuring these parameters are transmitted with Arduino Uno. Through a fast forest binary classifier, the obtained sensor information is examined. To decide whether the water samples are suitable for human consumption, a desktop application using the .NET framework is designed. Machine learning algorithms are also made in use to speculate the quality of water. All the results are stored in the database.

Krishna et al. in [12] addressed the layered architecture of the water monitoring system. Three layers are the data acquisition layer, data transmission, and monitoring. Tools and methodology used in the system are also mentioned. The desired sensor data can be derived from EXO Sonde and industrial multi-parameter sensor nodes. The obtained information is analyzed and transferred to a server on a layered architecture. The sensor node uses the file system to protect against data loss due to poor network. For communication between sensors and nodes, GSM is used. The information is stored and can be viewed on the web UI on the server.

Aditi Maulekhi et al. in [13] discussed problems faced in traditional water harvesting systems. They have proposed a system to reduce wastage of water as the water transferred process is simplified using various IoT sensors. Water quality

Table 1 Work carried out by other researchers

S. No.	Author	Year of publication	Work
1	Arjun Singh, Shashank Roy, Suresh Kumar Natarajan [1]	2017	The authors discussed that defining such a device will achieve maximum use and will be very effective in minimizing water wastage. They have observed real-life problems in schools, colleges, the private sector, and the apartment building where water management was not properly done. This system is also very simple and well-defined to implement. And by experimenting with this, we also got desired outcome we want and developed a well-defined and efficient water wastage management program
2	Raj Kumar Gupta, Karunesh Kumar Gupta, and Punit Khatri [6]	2019	They had discussed that monitoring water quality is important before use, and real-time monitoring can reduce the risk of human disease. Traditional methods for monitoring the water quality. This paper presents the design and development of a smart sensing device for measuring drinking quality: a keyboard, a touch screen, and a joystick. Only a software panel for the interactive user interface is being developed. Contrary to manual methods, the fuzzy approach employed in this paper is an effective way to detect drinking water quality in real time. They expected to introduce the Internet of Things (IoT) framework for online monitoring of drinking water quality in the distribution networks in the future

(continued)

Table 1 (continued)

S. No.	Author	Year of publication	Work
3	Saransh Shrivastava, Chinta Rohith Reddy, Guddu Kumar, Gowthamy J, Pijush Meher [2]	2018	The authors presented a prototype for water monitoring using IoT. The data obtained from all the sensors were used for analysis intent to solve water problems better. The data is sent via Wi-Fi module ESP8266 to the cloud server. Therefore, this program has been the toughest competitor in the real-time monitoring and management network and used to solve all water-related issues
4	K. A. Mamun, H. Haqva, F.R. Islam, A.N. Prasad [7]	2019	The authors are trying to talk about a method of intelligence control of the water quality. Four separate water sources were checked at regular periods over 12 h to verify the precision of the measurement methods. The relationship of temperature with conductivity and pH was observed in all the water samples. GSM is a technology which had been successfully deployed to submit an alarm focused on a particular user for urgent steps to ensure high quality of water based on the reference level. In addition, the level of references collected by miscellaneous water sources will be considered for creating classifiers which would be helpful in the context of Neural Network Analysis to conduct automated water analysis. This device has proven its importance by providing reliable and consistent data during test duration. With the additional functionality of integrating IoT systems for current time analyzing of water, this would be an outstanding competitor in real-time water monitoring solutions

(continued)

Table 1 (continued)

S. No.	Author	Year of publication	Work
5	SanjayAgrawal, M.B.Kawarkhe [8]	2019	They were presented that a device can automatically track water quality, is cost effective, and does not involve the person being physically present. This particular technique is helpful to prevent wasting an enormous quantity of water by unregulated usage of houses/offices and so on. Therefore, monitoring water quality is seen as highly cost-effective, efficient, and quick We can easily add a device for smart water to homes, businesses, classrooms, and wherever water tanks are used. By installing this device in intelligent building, they will be able to monitor and evaluate the residents water use habits, save a lot of water from waste, and provide good water quality from their side

is preserved. Using blockchain in the water industry, data about its consumption and usage of various IoT-enabled sensors by the users can be tracked. It consists of three tanks: tank one is used to store harvesting water, and tank two is where water is stored for daily purposes. Tank 3 is a central tank used to store water from multiple houses. All the information regarding water usage is stored in blockchain, together with data about individual water usage; an application is being used where users can enter their login credentials to access it.

Suma et al. in [5] discussed the rapidly growing population, which can lead to water scarcity. They proposed a system that includes the Smart Water Quality meter and Smart Water Quantity meter to ensure that any issues related to healthcare or unpredictable reasons can be avoided. Smart Water Quality meter examines five parameters which include temperature, dissolved oxygen, pH, conductivity, and turbidity test the purity of drinkable water received by the user. Water usage data are stored on the cloud and can be retrieved anytime through the web application.

The main motivation of the authors to design a Smart Water Quantity meter is to enable water conservation by observing the quantity of water used in a particular house. A monthly consumption bill for water consumed by a user is generated through a three-slab billing system. An alarming notification is delivered to the user and the authority when it crosses a threshold.

Rapelli et al. [14] proposed a device in such a manner that can tack the water level available in real time. The framework was introduced with an embedded system, and IoT is used for communication. The device is low in cost, and Arduino Uno will control the appropriate amount of water at a regular time. Flow control can be done through a solenoid valve when it crosses a particular limit. Turning on and off the solenoid valve allows the necessary flow rate to be managed for the water supply at a particular time. User can drop a message according to their demand and based on water available for storage, it will determine amount of water they will receive. According to the water utilized by the user, a bill is generated. In a situation when a user requires more water than usual. For checking user billing and monitoring, a mobile application can be installed. All the information is stored in the cloud.

Li et al. [15] facilitate the authority efficiently in preserving the better condition for portable water in a particular area. An accessible, user-friendly, and broad coverage wireless system to monitor water quality is being proposed. Based on Wi-Fi, multi sensors are also used to collect the data. Multi-sensors are integrated with temperature, pH, and TDS sensors. It consists of three components: wireless network, hardware, and software. Wireless network modules transfer the data in the wireless network component. The hardware component consists of various sensors that are connected to the microcontroller. In software components, various programming software is used to obtain the data. All the gathered information is visible on the website. It can calculate, gather, and water readings can be viewed. Promptly, data is transferred through the Wi-Fi module to the IoT platform. To avoid any data loss during the process, data collected by sensors is stored in real time on the IoT framework.

Vimalathithan et al. in [16] proposed a system for smart village projects to detect any chemical leakage and determine water quality and distribution patterns. LoRa WAN is one of the solutions when cellular network is not accessible such as in hilly areas. The LoRa mote with other sensors is mounted in water tanks in different locations, which comes under cooperate limit in a village. The device will continue to track the level and quality of water in all the tanks. It can be displayed in a particular place where there exists current water distribution structure and can be managed from a single location.

Lakshmi et al. in [17] introduced a system for smart cities integrated with fog and cloud computing. It will manage consumer usage, monitor water quality, prediction of faults, and water theft identification. Faultless delivery of water based on customer need and even health monitoring health of the underground surface of the pipeline network. The virtual tracking of the supply through a grid pipe is laid underground using wireless devices. The motive is to provide customer satisfaction with safe drinking water without affecting the city infrastructure, and to manage the faulty pipes by narrowing the failure location.

Sulaiman et al. in [18] proposed a water quality monitoring system prototype where they have taken water level as a primary parameter. It becomes a very crucial factor in areas that are vulnerable to floods. A water level sensor is used, and when it crosses a certain threshold, information is displayed on social media platforms. On the cloud, data is being stored. On the remote screen, water level readings can be visible. Ultrasonic and water sensors are used in the system implementation. Table 1 represents the extensive work done by other researchers in this field.

Akash et al. in [19], in their research paper, discussed implementing IoT to enhance the quality and safety of food. High food quality plays a crucial role in public well-being. IoT solutions have been adopted in multiple domains. Background work of five years has been considered on use cases of IoT for food management. Authors have contributed by mentioning the existing development and came across potential challenges and missing areas where further research is needed. This field's commonly used smart sensors are temperature, gas, and humidity. Food management through IoT is not so widespread, and communication mediums are Bluetooth, Wi-Fi, RFID, and ZigBee.

In their survey paper, Rubi Nurajni Kamal et al. [20] **aim to understand IoT's role** in food supply chain monitoring. Qualitative research methodology has been utilized where output obtains showed improvement in efficiency and effectiveness of existing solution when IoT is deployed. Collected data through sensors help in decision making and also assists in overcoming the challenges by converging different technologies. GPS connected to the food transport vehicle allows real-time monitoring of that vehicle by tracking the location and ensuring timely delivery of food. Food products currently present in the warehouse can be identified with IoT. Table 1 reviews other works carried out by another researcher. This review paper examines a generic case for the smart water monitoring system. This provides a variety of possible benefits for water services.

The current water treatment system cannot determine the containments, including certain chemicals. The management system is unsafe when using a conventional water quality control system [21]. Chlorinating is being used to prevent micro-organisms. But drinking it in excess can lead to various diseases such as cancer. In addition, there is no specific tool that can measure all the potential parameters of water. Therefore, implementing a real-time online water quality monitoring system would help control waterborne diseases.

3 The Proposed Water Monitoring System

Through this, we present the concept of tracking the quality and quantity of water using IoT. The device comprises multiple forms of sensors such as ultrasonic sensor, pH, water flow sensor, and turbidity sensor. It also includes a microcontroller and Arduino. The system's main processor is Arduino, responsible for controlling and processing the sensor-produced data [22]. To transmit the data on the internet through the cloud, we connected Arduino to the Wi-Fi module. In a given time, the sensor used for water flow calculates the amount of water flows via the pipe, and then this data is transferred to the cloud for analysis and storage reasons.

We can determine the water's level when the flow exceeds a particular point through ultrasonic sensors. At that time, we can halt the flow of water naturally with the help of Arduino to close the water flow inside the pipe. The rest of the sensors, such as turbidity, pH, and temperature, calculate the water standard.

3.1 Components of the System

Figure 1 represents a block diagram of the system. It has several components which help to build a water monitoring system. The essential element in the smart home automation system is described in the following sub-sections.

3.1.1 Arduino Uno

Arduino is ATmega328 based on a microcontroller board. This component has six analog inputs, a USB connection, an ICSP header, 14 digital output and input pins, power jack, 16 MHz quartz crystal, and a reset button.

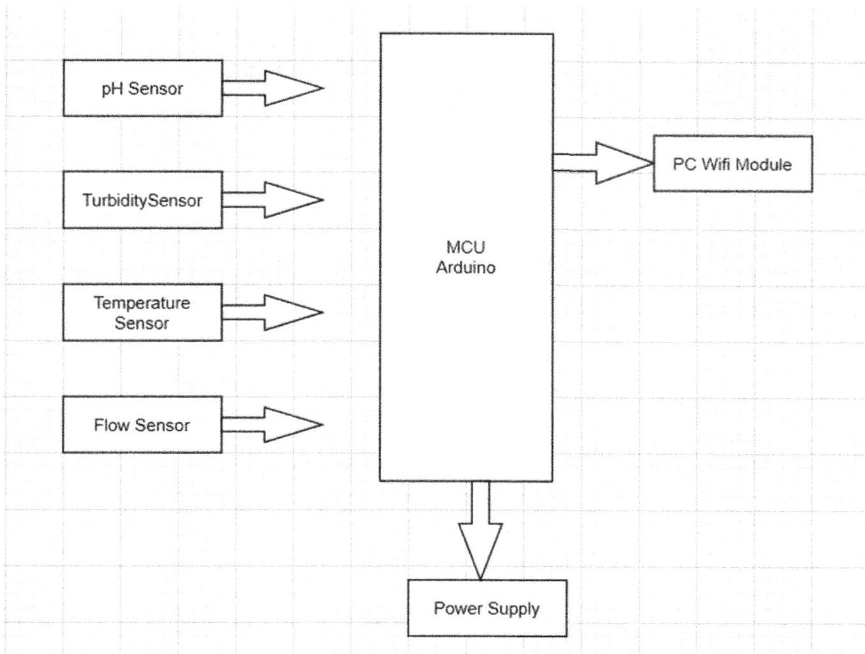

Fig. 1 Block diagram of the system

3.1.2 Server Based on the Cloud

To store data produced by home installed sensors, cloud goes around the database as a service. Through this service, we can send email updates to the client about various circumstances at home.

3.1.3 Wi-Fi Module

The ESP8266 conduct an application or upload the rest of the Wi-Fi network function from the separate processor of an application. This Wi-Fi module allows access to your Wi-Fi network to any controller through a self SOC with a built-in TCP/IP protocol [23]. Every module of ESP8266 comes well programmed with a firmware set of AT commands. It is a very cost-efficient module.

3.1.4 Sensors

An electronic system that senses the physical condition and reacts to some form of information. pH sensor, turbidity sensor, temperature sensor, and an ultrasonic sensor that identifies changes in atmospheric disturbances are used in this specific form of sensor [22].

3.1.5 Flow Sensor

This component includes the rotor, sensor, and plastic frame. As liquid flows via a pipe, a pinwheel rotor rotates. Its speed is direct proportion to flow rate [22]. Hall Effect sensor provides a pulse for every pinwheel rotor revolution.

3.1.6 Ultrasonic Sensor

This sensor antenna translates signal turn into ultrasound; recipients translate ultrasound into electrical signal and transceivers can transfer ultrasound and accept it. This provides us with how to calculate the level of water.

4 Research Methodology

1. With the help of Arduino controller and different sensors (temperature, conductivity, turbidity and pH) linked to Internet of Things.
2. In this analog signal, these four sensors store the data. To transfer the information of four signals into the digital format, we have used an ADC converter.

3. The digital signals are transferred to the Arduino in conjunction with the board.
4. The Arduino microcontroller would analyze it and perform digital analysis.
5. The Wi-Fi module will submit factors for water quality to cell phones/ personal computers via notification for contacting the next channel, which will be seen on LCD. The microcontroller acknowledges the data and integrates the data obtained via a Wi-Fi module from sensors in the browser.
6. The data which is obtained can be viewed in real time.

5 IoT in Water Industry

The Internet of Things is a collection of inter-connected devices, vehicles, and appliances capable of collecting and sharing information without human interaction [1]. IoT is a platform in which embedded devices are connected to the internet so that they can gather and share data. This encourages device interaction, collaboration and benefits from each other experiences just like human beings do. We can connect any device to IoT, provided it has sensors and an Internet connection. Specific IoT-equipped devices, to name a few, include laptops, smartphones, refrigerators, coffee machines, Google Home, Fitbits, and Apple Watches [7].

The Web is a living, ever-changing and developing thing. New applications and companies are continually developed. Besides a changing Internet, technology changes the landscape too. Broadband connectivity is getting cheaper and more ubiquitous; devices with a wide range of onboard sensors are becoming more powerful and smaller. The emergence of more interacting apps leads to a new paradigm: The Internet of Things. An Internet growth guides the Internet of Things through integrating physical objects and the ability to provide the environment with smarter services as more data becomes obtainable [24]. Figure 2 depicts the applications of IoT in the water industry.

5.1 Increasing Agricultural Efficiency

IoT is transforming the world of agriculture and enabling farmers to overcome the immense hardships they face. It must address the growing water shortage and limited land availability, thus serving a world population with increasing consumption needs. Irrigation plays an important role in agriculture, but with the passing year, we require more water usage than rainfall. Farmers must find ways to conserve water while producing the maximum yield becomes important [1]. We can tackle the abovementioned situation and increase agriculture efficiency, productivity, and cost effectiveness [8].

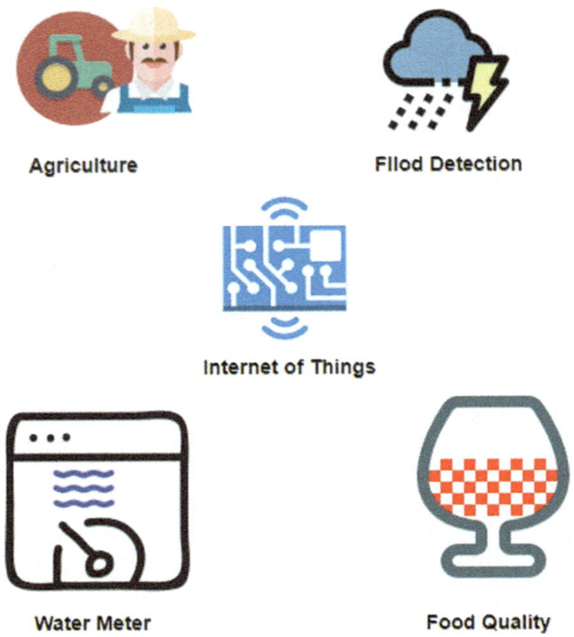

Fig. 2 IoT in water industry

5.2 *Water Leakage Detection*

Water leakage usually goes unnoticed or will only be alerted to after the incident. Thus, due to unnecessary irrigation, a large amount of water is wasted (only 70% of water distributed is used by farming. There is increasing demand for innovative solutions to make further effective use of available water supplies, enhance the drinking quality of water and boost water management planning [8]. IoT provides accurate water resource information monitoring, enabling water industries to be controlled efficiently and optimally. The smart water monitoring system will bring about a rapid and great improvement in water supply cost and efficiency, especially in farming and urban areas [25].

5.3 *River Flood Detection*

River flood detection is critical to mitigating the cost of flood damage as the expense caused by a flood is closely associated with the warning period provided before the flood incident. When the level of water rises above the crucial point, the device gives an alarm to the user. That device will alert by standby level, the level of evacuation, and the level of danger. By using IoT, we will get details regarding the actual water level like lakes and drainage in the specific region [23].

5.4 Water Meter

The IoT-powered meter comprises intelligent water meters that transmit regular testing to a receiver network sharing data between sensor and CRM and removing the need for an individual meter. Driven by IoT smart water meters, it can allow water suppliers to monitor utility water use efficiently. The need for IoT-enabled smart water meters is steadily growing due to their numerous real-time simulations, leakage identification and machine-to-machine interaction functionality [23].

5.5 Water Supply Management

For real-time monitoring of the amount of water consumed by the end user, utilities use smart meters [26]. Through smart metering and level monitoring helps utilities to define and manage the requirement of supply in a particular location. The data obtained can estimate the need for water in an area, and water distribution can be done accordingly. Water stored in overhead tanks can also be tracked to make it much more efficient.

5.6 Wastewater Treatment

IoT may be used to classify the pollutant mixed wastewater in wastewater treatment and then move it through various treatment methods to clean it cost-effectively. To achieve this, we can use built-in analytical features of IoT systems to enhance water treatment efficiency and improve water-struck areas' conditions.

6 New Innovations

Smart water technology today provides clarity and water control to the entire water supply chain, from freshwater to storing and treating wastewater. Here are the appliances, systems, and software tools that allow the benefits of smart water and help optimize water treatment, processing, distribution, and usage.

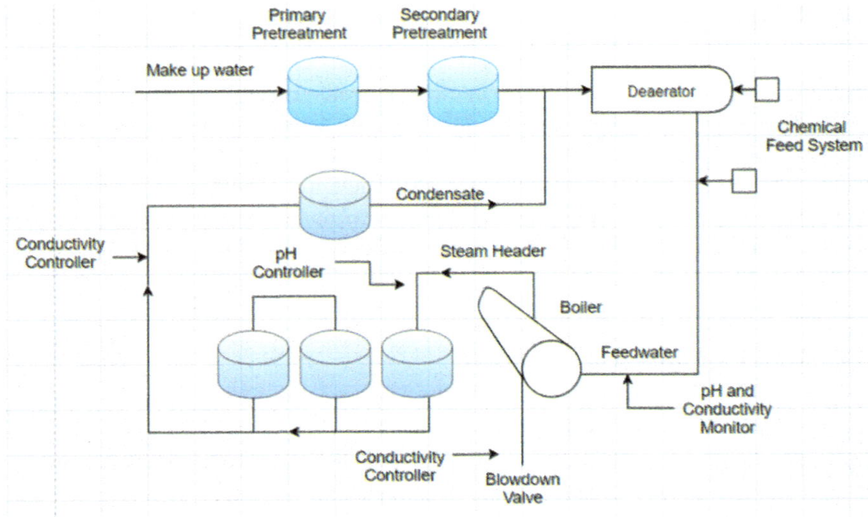

Fig. 3 Sensors in water industry

6.1 *Sensors*

Smart sensors are mounted at different positions in the water system. Once these sensors obtain data, they are sent back to the intelligent monitoring system. Figure 3 presents different sensors used in the water industry.

Due to their great versatility and intent, sensors have wide applications in smart water management. Sensors calculate in a very simple water supply chain: the consistency of the raw catchment water, the chemical composition of the water after treatment and wastewater, etc. [27]. Changing amounts in the storage tank, the pressure on the pipes in the distribution pipeline. Wear the machinery, equipment that filters and distributes water to end users, and much more.

6.2 *Smart Meters and Surveillance Devices*

Smart meters and analytical platforms allow real-time measurement of water use, help identify excessive use and waste point, correct use trends, and make predictions for potential use. This water management system is useful for the management of water production as well as distribution, especially for bulk households. Both can correct their water consumption habits and achieve sustainability and budgeting targets using smart meters and market monitoring devices [28].

6.3 Automated Delivery Systems

A smart distribution system uses environmental sensors and predefined or self-learning algorithms to provide automatic water supply to the endpoint [3]. For example, in the case of smart irrigation, sprinkles provide just enough water depending on soil moisture readings, air moisture, and sensors for crop conditions,

7 Applications of Water Monitoring for Food Quality

7.1 Smart Irrigation

Water quality is important to crop growth. In general, poor quality water can slow the growth of plants and harm their appearance. To add on, some plants grow faster in irrigation water when particular characteristics are available; Similar case the, conversely, certain factors can impede plant development [29]. For example, water with too high a pH level (a predictor). Another problem concerns resistance to the water. The plants used to produce nutrients and consume ionic salts, but some plants can manage high salinity, whereas others cannot. The water of irrigation differs in salinity; salinity is typically studied as water's electrical conductivity as salts that are dissolved conduct electricity inside the water. The higher its conductivity, the greater the concentration of salts [28].

7.2 Drinking Water

Most cities in the United States are actively using water quality control equipment to determine the condition of their water sources. The capacity for oxidation reduction (ORP) effectively tests water's ability to break down contaminants; hence some municipalities use ORP sensors to measure the oxygen dissolved in water. Drinking water is usually analyzed for the presence of bacteria, so testing the sterility of water is critical—can it sustain life [30]?

7.3 Aquaculture

Aquaculture, which involves aquatic plants and fish farming, demands water quality, increasing profitability, and sustaining production. Appropriate quality of water varies by species, so criteria differ across the industries. Many companies are also using control on water quality devices to measure conductivity & temperature, which include two of the several variables that will influence aquatic life. This widely monitored water quality parameters among farming systems, including weather, level of pH, H_2O, Alkalinity, ammonia, nitrites, and hardness [30]. CO_2, salinity, and chlorides may also be tracked depending on the method.

7.4 Detection of Water Loss

Roughly most high-rising buildings have a hidden pipes network carrying water all over the building. In the rest of the area, leaks in these pipes are hard to find, and even when they are discovered it is usually after a large amount of water has been wasted already [9]. Leakage of pipes is also a factor for water pollution in high amounts in manufacturing and the community system of water [28]. Water flow sensors monitor water flow through a pipe and its rate of change. Flow sensors are helpful in manufacturing installations for calculating the volumetric flow rate of any stream, gas, or liquid. Meters of the pulse is useful in understanding the normal flow rate & thus provide a cut of detecting changes in normal velocity.

7.5 Low Wastage of Food

IoT-based inventory management will allow for early detection of expiry dates, and stocks will be prepared [9]. Thus, food wastage would be brought down due to the absence of spoiled food. Smart transportation and advanced storage facilities enable products to be kept in an ideal situation based on product demands. In this way, they don't get spoiled. With digital counterparts, products can update their quality by themselves, allowing employees to detect food that might expire soon [12]. IoT also provides decision-making with the help of obtained data in real time.

7.6 Controlling Food Quality

Remote food quality monitoring is made possible with IoT-based smart refrigerators and transport vehicles where temperature can be set based on the food requirement. Temperature is an important parameter to maintaining the good quality of food and temperature varies for different food requirements [11]. Any update in the storage facility can be identified much quicker, which will help cover the challenges more easily. Advance transportation maintains the food quality consistently.

7.7 *Water Parameters*

7.7.1 Level of pH

The pH is an indicator of the acidic solution or alkaline. The pH is a scale logarithmic with a range from 0 to 14, and 7 is the neutral point. Values 7 above reflect a basic solution, and values before 7 reflect that solution is acidic. Much marine life requires a pH of 6.5–9.0. Something beyond this ideal range is lethal to the aquatic environment [7]. High pH values also maximize the viscosity of compounds and elements, which make them toxic and thus influence them to be absorbed by aquatic life in most cases. The temperature has an opposite relation with the pH level as temperature increases and decreases pH levels and inversely [5].

7.7.2 Temperature

It is important to report temperature along with the rest of the parameters because it benefits the parameters being measured in the behavioral analysis. Regarding this, high temperature is of significant concern for pacific island climates. In temperature-relationship theory, conductivity and level of pH have an adverse association with significant temperature change [16].

7.7.3 Oxidation Possible Reduction

Oxidation–Reduction Potential (ORP) includes the calculation of oxidizing strength of the solution. This can be defined in simple terms as the capacity of the ability of a solution to sanitize itself. High Oxidation Reduction Potential values will suggest that there will be more oxidizers [12]. Similarly, lower ORP is equal to more reducers current. It is known that 100 mV to 200 mV would generally be a good value for aquatic life [7]. Something beyond those limits is a case to be examined. A similar case can be said for water from the tap with very high ORP levels, usually in the high 600 mV, because of disinfectants.

7.8 *Conductivity*

Conductivity refers to the strength which is ionic of the solution. On the other side, it refers to the solution's capacity to conduct electricity using the measuring unit as a micro-Siemens per centimeter. When the ions which are dissolved in the water increase their conductivity. The tap water conductivity is really small at approximately 200 cm [12]. The predicted range for sea water is around 55,000–60,000 cm due to high ionic data content. Any more rise in this rate may suggest a contaminated environment, including sewage leakage or water flooding waste of chemicals [7].

8 Case Studies

8.1 Smart Sensors for Monitoring the Water Quality of Fish Farms

Plug and sense sensors by Libelium to monitor the productivity of farms by reducing cost and number of causalities. Sensors are responsible for measuring: Nitrite, pH levels, water temperature, dissolved Oxygen, and Ammonium. Ensures real-time tracking of fish ponds for uttermost efficiency [31]. The study showed that deploying smart sensors in fish farms has reduced fatalities, expanded productivity, and minimized production costs [17].

8.2 Cannon Artes Solution for Improving Water Quality in the Food and Beverage Sector

Water is a crucial part of food production. Ultrafiltration and microfiltration eliminate bacteria and remove solids. Reverse Osmosis enables deionization and allows permeate reusability for generating steam. It also acts as an ingredient while preparing food. Proper water treatment should be utilized in different applications [32].

8.3 Manzana Achieves Real-Time Traceability, Which Enhances Food Processing

Manzana is an apple cannery in California looking for an innovative solution to produce real-time visibility and revolutionize the inventory and supply chain. Microsoft Dynamics was deployed to achieve high ROI. Capabilities of inventory monitoring are also increased [17].

8.4 Jasper Wyman and Son Boosted Traceability

Jasper Wyman & Son headquarters in Milbridge is engaged in the frozen food field and sells its products in three different modes: food service, retail, and ingredient. Advance data recording enables organizations with food inspection. Organizations share the data online when consumers ask for product details with lot numbers [18]. All the details are safely stored in Dynamics NAV.

9 Conclusion and Future Scopes

The system is low in cost, and real-time monitoring can be done. It does not require humans on board. The system can be easily implemented where water tanks are used, such as in residential areas, offices, and universities. Huge water waste can be brought under control at offices/schools.

Monitoring and control systems play a significant role in the day-to-day service and maintenance of water supply and sewage systems. Smart water networks based on a combination of the Internet of Things and Artificial Intelligence technology will help avoid such forecasts and reverse the harm that has already been done by the unwise use of water resources [15]. When implementing IoT technology, water data needs to be transmitted over a longer distance, wirelessly and uninterrupted, to a central dashboard to evaluate and track the water network. In the broad context, smart water means water storage and delivery, thus preserving its consistency.

References

1. S. Wadekar, V. Vakare, R. Prajapati, S. Yadav, V. Yadav, Smart water management using IOT. In *2016 5th International Conference on Wireless Networks and Embedded Systems (WECON)* (IEEE, 2016), pp. 1–4
2. J. Gowthamy, C.R. Reddy, P. Meher, S. Shrivastava, G. Kumar, Smart water monitoring system using IoT. Int. Res. J. Eng. Technol. **5**(10), 1170–1173 (2018)
3. T. Kumura, N. Suzuki, M. Takahashi, S. Tominaga, S. Morioka, S. Ivan, Smart water management technology with intelligent sensing and ICT for the integrated water systems. Special Issue on Solutions for Society-Creating a Safer and More Secure Society **9**(1) (2015)
4. V. Choudhary, S. Tanwar, A. Rana, Demystifying security and applications of Internet of Things. In *2021 9th International Conference on Reliability, Infocom Technologies and Optimization (Trends and Future Directions) (ICRITO)* (IEEE, 2021), pp. 1–5
5. M.K. Jha, R.K. Sah, M.S. Rashmitha, R. Sinha, B. Sujatha, K.V. Suma, Smart water monitoring system for real-time water quality and usage monitoring. In *2018 International Conference on Inventive Research in Computing Applications (ICIRCA)* (IEEE, 2018), pp. 617–621
6. P. Khatri, K.K. Gupta, R.K.Gupta, Smart water quality monitoring system for distribution networks. In *Proceedings of International Conference on Sustainable Computing in Science, Technology and Management (SUSCOM)* (Amity University Rajasthan, Jaipur, India. 2019)
7. A.N. Prasad, K. Al Mamun, F.R. Islam, H. Haqva, Smart water quality monitoring system. In *2015 2nd Asia-Pacific World Congress on Computer Science and Engineering (APWC on CSE)* (IEEE, 2015), pp. 1–6
8. M.B. Kawarkhe, S. Agrawal, Smart Water monitoring system using IoT at home. J Computing **21**(1), 14–19 (2019)
9. A. Meshal, M.N. Mikhael, H.A. Mansour, Water quality monitoring system of the Nile River. J. Homepage: http://erj.bu.edu.eg 1, **43**, 100–104 (2020)
10. S. Pasika, S.T. Gandla, Smart water quality monitoring system with cost-effective using IoT. *Heliyon* 6, no. 7 (2020): e04096

11. M. Mukta, S. Islam, S.D. Barman, A.W. Reza, M.S.H. Khan, Iot based smart water quality monitoring system. In *2019 IEEE 4th International Conference on Computer and Communication Systems (ICCCS)* (IEEE, 2019), pp. 669–673
12. S. Krishna, T.V. Sarath, M.S. Kumaraswamy, V. Nair, IoT based water parameter monitoring system. In *2020 5th International Conference on Communication and Electronics Systems (ICCES)* (IEEE, 2020), pp. 1299–1303
13. A. Maulekhi, K.A. Patani, V. Sangani, M.V. Pawar, The blockchain-based smart water harvesting system. In *2020 International Conference on Smart Electronics and Communication (ICOSEC)* (IEEE, 2020), pp. 771–776
14. N. Rapelli, A. Myakal, V. Kota, P.R. Rajarapollu, IOT based smart water management, monitoring and distribution system for an apartment. In *2019 International Conference on Intelligent Computing and Control Systems (ICCS)* (IEEE, 2019), pp. 440–443
15. L.Y. Li, H. Jaafar, N.H. Ramli, Preliminary study of water quality monitoring based on WSN technology. In *2018 International Conference on Computational Approach in Smart Systems Design and Applications (ICASSDA)* (IEEE, 2018), pp. 1–7
16. A.M. Manoharan, V. Rathinasabapathy, Smart water quality monitoring and metering using Lora for smart villages. In *2018 2nd International Conference on Smart Grid and Smart Cities (ICSGSC)* (IEEE, 2018), pp. 57–61
17. L.K. Narayanan, S. Sankaranarayanan, IoT enabled smart water distribution and underground pipe health monitoring architecture for smart cities. In *2019 IEEE 5th International Conference for Convergence in Technology (I2CT)* (IEEE, 2019), pp. 1–7
18. T. Perumal, M.N. Sulaiman, C.Y. Leong, Internet of Things (IoT) enabled water monitoring system. In *2015 IEEE 4th Global Conference on Consumer Electronics (GCCE)* (IEEE, 2015) pp. 86–87
19. R.M. Dias, G. Marques, A.K. Bhoi, Internet of Things for enhanced food safety and quality assurance: a literature review. Adv. Electron. Commun. Comput., 653–663 (2021)
20. H. Maulana, S.L.B. Ginting, P. Aryan, M.R. Fadillah, R.N. Kamal, Utilization of Internet of Things on food supply chains in food industry. Int. J. Inf. Inf. Syst. Comput. Eng. (INJIISCOM), **2**(1), 103–112
21. M. Bolbol, H. Muhammetoğlu, A. Muhammetoğlu, Smart water management for smart cities. In *3rd International Students Science Congress*, p. 229
22. S. Geetha, S.J.S.W. Gouthami, Internet of things enabled real time water quality monitoring system. Smart Water **2**(1), 1–19 (2016)
23. K.M. Shahanas, P.B. Sivakumar, Framework for a smart water management system in the context of smart city initiatives in India. Procedia Comput. Sci. **92**, 142–147 (2016)
24. C.Z. Myint, L. Gopal, Y.L. Aung, Reconfigurable smart water quality monitoring system in IoT environment. In *2017 IEEE/ACIS 16th International Conference on Computer and Information Science (ICIS)* (IEEE, 2017), pp. 435–440
25. S. Nikoletseas, T.P. Raptis, A. Souroulagkas, D. Tsolovos, Wireless power transfer protocols in sensor networks: experiments and simulations. J. Sens. Actuator Netw. **6**(2), 4 (2017)
26. B. Vidya, K. Poonam, G. Priyanka, D. Gaurav, Prof. A.S. Chandgude, Water level monitoring system in real time mode using WSN. Int. J. Emerg. Technol. Adv. Eng. (ISSN 2250–2459, ISO 9001: 2008 Certified Journal, **6**(9) (2016)
27. M. Manju, V. Karthik, S. Hariharan, B. Sreekar, Real time monitoring of the environmental parameters of an aquaponic system based on Internet of Things. In *2017 Third International Conference on Science Technology Engineering & Management (ICONSTEM)* (IEEE, 2017), pp. 943–948
28. https://www.digiteum.com/smart-water-management-iot. Accessed 13 Jan 2022
29. A. Faustine, A.N. Mvuma, H.J. Mongi, M.C. Gabriel, A.J. Tenge, S.B. Kucel, *Wireless Sensor Networks for Water Quality Monitoring and Control within Lake Victoria Basin* (2014)
30. P. Sachdeva, S. Katchii, A review paper on raspberry pi. Int. J. Curr. Eng. Technol. **4**(6), 3818–3819 (2014)

31. T.A. Al-Janabi, H.S. Al-Raweshidy, Optimised clustering algorithm-based centralised architecture for load balancing in IoT network. In *2017 International Symposium on Wireless Communication Systems (ISWCS)* (IEEE, 2017), pp. 269–274
32. https://www.cannonartes.com/markets/food-and-beverage/. Accessed on June 2022

An Approach to Optimizing Food Quality Prediction Throughout Machine Learning

Mohammed M. Ahmed and Aboul Ella Hassanien

Abstract Everything in our life requires real-time monitoring and control, especially when preventing food from spoiling. Keeping the food from spoiling bits helps keep food from spoiling and reduces incidental losses in the business. An Internet of Things (IoT)-based framework for food monitoring is proposed in this paper to protect food from spoilage caused by changes in environmental conditions during storage. In the current scenario, the prediction was made using recorded sensed data, and a detailed analysis was performed to identify the factors causing the food to spoil. This chapter proposes an automated controlling mechanism for environmental parameters using Adaptive Random Forest prediction and IoT. Environmental parameters such as temperature, humidity, moisture, light, and so on, which affect the nutritional value of food, are considered in the proposed work. All sensed data would be stored on the cloud, and analysis will be performed to predict the environmental condition at the storage location to avoid food spoilage by changing to the appropriate environmental condition. In the proposed work, the Random Forest Optimizer method is used to predict the state of food if it is fresh or not at the location where food is stored, harmful changes are monitored, and action is taken to provide appropriate conditions at the stored location. Random Forest optimizer has scored the highest accuracy of 94.01%.

Keywords Food quality prediction · Random forest optimizer · Prediction

M. M. Ahmed (✉)
Faculty of Computers and AI, University of Sadat City, Sadat City, Egypt
e-mail: mohammed.mostafa@fcai.usc.edu.eg

A. E. Hassanien
Faculty of Computers and AI, Cairo University, Cairo, Egypt

M. M. Ahmed · A. E. Hassanien
Scientific Research Group in Egypt (SRGE) Egypt, Cairo, Egypt

A. E. Hassanien and M. Soliman (eds.), *Artificial Intelligence: A Real Opportunity in the Food Industry*, Studies in Computational Intelligence 1000,
https://doi.org/10.1007/978-3-031-13702-0_9

1 Introduction

Food poisoning is caused by changes in environmental parameters such as temperature and humidity. Food pollution can occur during the production, transportation, and storage phases; however, pollution is frequently induced by inefficient food handling in unsuitable environmental conditions during transportation and storage [1–6]. As a result, temperature and humidity must be regulated throughout food transportation and storage [7].

When a food product degrades, it becomes unsafe for the client to consume. The components of such a technique are influenced by various external factors, such as the type of food item and how it is packaged and processed [8]. One-third of the world's melded is lost every year due to meal waste. Machine learning has already been utilized to forecast the freshness of fish [9] and puffed snacks [10], as well as evaluate the quality of olive oil [11] using sensory assessment learning findings.

Various techniques, including K-Nearest Neighbor, support vector machine, Artificial Neural Network, and Convolutional Neural Network, have been used in computer vision systems for categorization in food quality assessment [12]. The SVM model is a machine learning method that can be used for regression and classification. The SVM classifier has been applied in various fields, including plant disease detection and classification [13] and tomato maturity assessment [14]. An artificial neural network (ANN) is a computational modeling tool widely used for simulating complex real-world issues [15]. As human senses [16] demonstrated, the SVM model and ANN are adaptive and have nonlinear learning capabilities. The Internet of Things (IoT) requires a few key components to facilitate communication between devices and things. Objects must be equipped with auto-identified technology, such as a radio frequency identification (RFID) tag, to be uniquely recognized. RFID is a well-known and widely utilized auto-identification technology that can be used in traceability systems to track and monitor things, particularly in various supply chains. In PFSC, RFID was effectively applied and showed considerable results for the traceability system [17].

The contribution of this paper is to predict food quality, fresh or non-fresh, via a random forest optimizer. When fruits like bananas are stacked together in a warehouse, the ethylene gas produced by the first banana triggers a ripening wave. This mechanism speeds up the ripening of other bananas, resulting in food products being overripe sooner than intended. To reduce food waste, continual monitoring is essential.

The chapter is structured as follows. Section 2 goes over related works. Section 3 contains the key concepts for Random Forest. Section 4 discusses the proposed system approach for predicting food quality from a dataset. Section 5 discusses the experiment results. Section 6 includes the conclusion of the work.

2 Literature Review

Venkatesh et al. [18], Bluetooth and IoT-based systems have numerous advantages in commercial domains such as industries, medical, cosmetics, agriculture, production, security, and research. Why not apply similar techniques to food safety in the food industry? The proposed method is based on environmental factors such as temperature, humidity, and gases released by the meal. According to Sivakumar and Janaki [19], the global food passing committee recommends a 25 percent calorie intensity in foods. Food monitoring agencies such as the United Nations Food and Agriculture Organizations also concentrate on similar food quality measurement techniques.

Al-Marakeby et al. [20] presented a vision-related categorizing system to improve the prominence of food goods. The classification process is based on retaining the image of the fruit or product and reviewing it to identify any defective items. An image processing-based apple fruit disease classification system was presented and verified by Dubey and Jalal [21]. The method has given consisted of four steps. The region of interest was extracted using the K-means clustering-based defect segmentation method in the first stage. The second step used the segmented apple illnesses to extract state-of-the-art color, texture, and shape-based attributes. Yu et al. [22] created a Hybrid Adaptive framework (HAEL) that integrates adaptive processes like "base classifier competition" and "classifier ensemble interaction" for studying favorable variable datasets at the same time. Huang et al. [23] proposed a two-level representation for "Fast Image Processing via Random Forests" (FIRF) that delivers high accuracy while using minimal computational power.

Maksimovic et al. [24] argue that human health should take precedence over all other aspects of the cosmos. Because food quality is directly linked to one's health, we must be aware of food safety and quality. Pabler in [25] proposed a hierarchical IoT-based architecture for non-invasive monitoring of human food intake behavior for long-term dietary restrictions utilizing chewing and swallowing sound sensors. Consumer perceptions of food freshness have been studied using a broader definition of freshness [26]. Several research studies have combined consumer ratings and objective metrics [27].

Megalingam et al., present a novel technique for identifying food degeneration by combining image classification and AI [28]. Green et al. demonstrated an automatic method of four gas sensors composed of functionalized single-walled carbon nanotubes (f-SWNTs) and polymer nanocomposites to become a simple monitoring device [29]. Ahmed et al. [30] optimized machine learning model like K-Nearest-Neighbor (KNN) to increase wireless sensor network lifetime. Machine learning clustering improves the energy selection process when CHs are based on a fitness function that takes into account the residual energy of the node and the total energy of nearby nodes [31].

3 Preliminaries

3.1 Random Forest (RF)

Breiman's Random Forest (RF) [32] is an ensemble algorithm widely used in batch or non-stream learning for regression, classification, and other tasks. The algorithm generates some decision trees before combining them into a single model. The final prediction made by the algorithm for classification problems is based on the majority of votes from all of the trees, where each tree makes a class prediction.

Bagging [33] is an ensemble algorithm that generates a specified number of datasets, each with N records, from a given training dataset. The datasets are created by picking records at random and replacing them. As a result, each record from the original dataset can be replicated K times in each bootstrapped sample. While growing the trees and splitting each node, a random selection of features is used. The whole set of features is reviewed at each node in the typical decision tree algorithm to choose the essential feature to split. In the random forest, however, each node is split using a random subset of $m<M$ features, where M is the total number of features. The RF technique is more powerful than simple ensembles of trees and less prone to overfitting if the forest contains enough trees because it uses bagging and random feature selection. When tested on different data sets, it was also proven that the Random Forest method is fast to train and effective.

The random forest approach generally uses both implicit and explicit diversities, and its effectiveness is determined by the accuracy and variety of the classifiers [34]. Furthermore, research has shown that choosing the number of decision trees in a random forest is critical and that adding many classifiers to the random forest does not improve accuracy. As a result, studies in the literature are being conducted to reduce the number of classifiers and identify the best subset of classifiers [35].

3.2 Ant Colony Optimization (ACO)

The Ant Colony Optimization approach uses iterations to find the optimal feature subset [36]. The primary goal of the Ant Colony Optimization approach is to reduce redundancy. Each ant chooses the features with the least similarity to the previously chosen features. As a result, if a feature has the least similarity to pheromones, the likelihood of its selection by other ants will increase in following rounds. Finally, because the characteristics are comparable, the selected major features will have high pheromone values. As a result, the ACO technique selects the best features with the least amount of redundancy. The feature selection problem is depicted in Fig. 1.

The steps for selecting ACO features are shown. The search space must be described in a proper form for ACO before the feature selection approach can begin with this methodology. As a result, the search space is defined as a fully connected

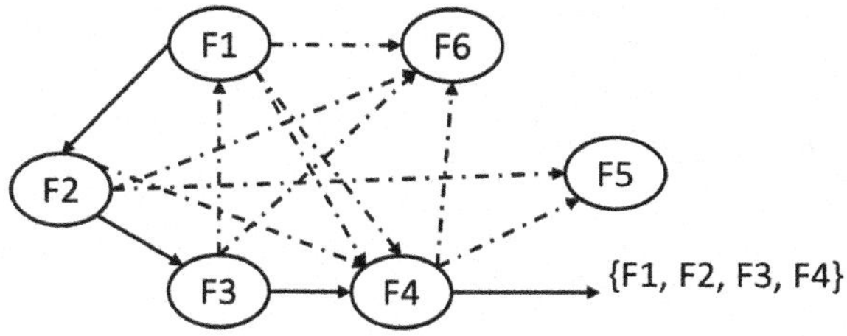

Fig. 1 The feature selection problem representation

undirected weighted graph, $G = (F, E)$ where $F = \{F1, F2, \ldots, Fn\}$ represents a collection of all features in that each feature marks a node in the graph, and $E = \{(Fi, Fj): Fi, Fj \in F\}$ represents the node edge.

4 The Proposed Approach

The major goal of this study is to predict food quality (fresh or not fresh) using the classifier Random Forest, so it needs to increase the random forest algorithm's performance by generating accurate and diverse classifiers using various well-performing training sets. Because accuracy and variety are trade-offs, these two criteria are maximized simultaneously to build accurate and diverse decision trees and identify the (near) ideal number of trees.

ACO is employed as a metaheuristic algorithm in this work to search the space of all potential subsets of features on the training portion of the original dataset. ACO generates candidate feature subsets that are assessed using an inductive algorithm and some evaluation function. Random Forest classifier is used as an inductive algorithm and as an evaluation function. The inductive method is trained and validated utilizing a cross validation approach to increase the reliability of the evaluation process. The Random Forest model is trained in the second step using the best features subset obtained in the first stage, and then it is tested on another unrepresented part of the dataset (testing part) to obtain the final evaluation. Figure 2 depicts all the details of the proposed approach.

Table 1 depicts a sample for a dataset for IoT sensor data generated from the food environment. We can see the data frame consists of 3 features: gas concentration, temperature value, and humidity values are the input features, while in the last column fresh state is the target value that our model must predict.

4.1 Data Collection

Data collection from IoT devices that monitor the environment The temperature and humidity sensor ensured regular monitoring of the room conditions and took both

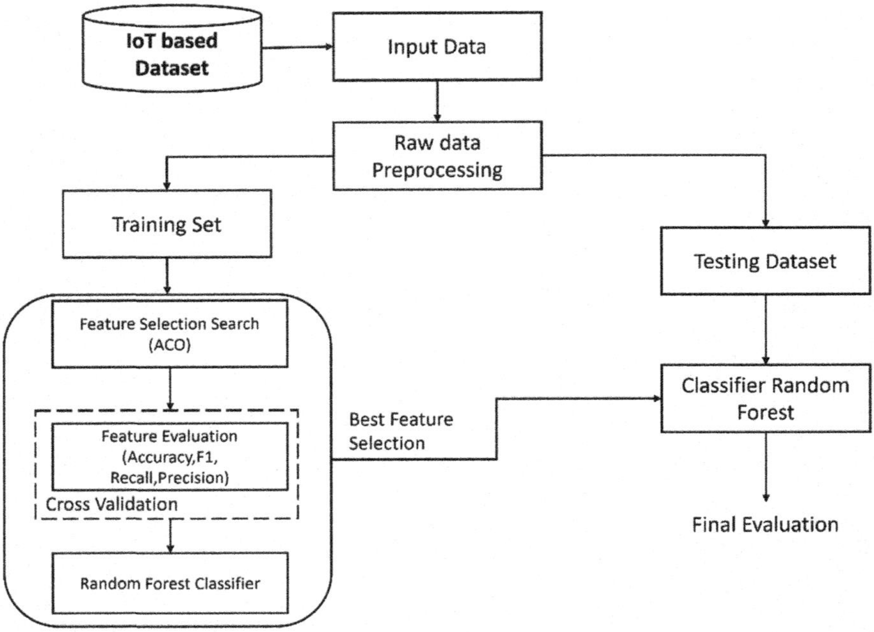

Fig. 2 Flowchart of the proposed approach

Table 1 Sample dataset (15 records)

Index	Gas	Temperature (ppm)	Humidity	Fresh
0	0.072333	0.430891	0.760454	0
1	0.072333	0.601714	0.564117	0
2	0.072333	0.573267	0.567956	0
3	0.072333	0.397121	0.671301	0
4	0.072333	0.59301	0.798849	0
5	0.072333	0.22218	0.721367	0
6	0.072333	0.543214	0.657411	0
7	0.072333	0.444765	0.754142	0
8	0.072333	0.49504	0.808633	0
9	0.072333	0.591021	0.658118	0
10	0.072333	0.493202	0.749647	0
11	0.072333	0.493925582	0.759358	0
12	0.072333	0.494649164	0.76907	0
13	0.072333	0.495372745	0.778781	0
14	0.072333	0.496096327	0.788492	0

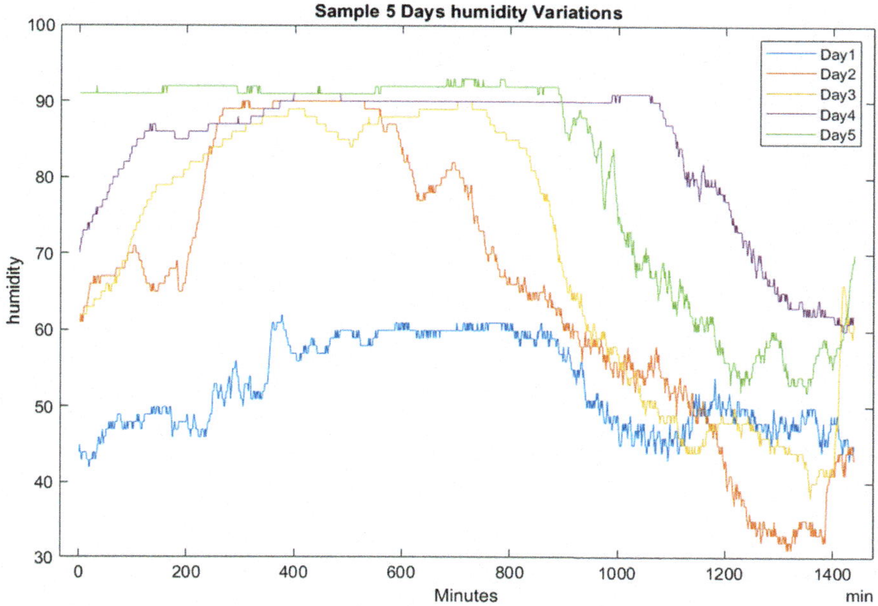

Fig. 3 Sample for humidity dataset

humidity and temperature Figs. 3 and 4, respectively into account in real-time. In this section, the data extracted from the sensors are collected from the temperature and humidity affecting the food. Three thousand samples were collected at a pace of 30 s each sample.

4.2 Data Preparation

Data Pre-processing is an essential phase in the information mining process since it deals with modifying or removing information before it is used. Information pre-handling in AI is the process of planning (organizing and cleaning) raw data to be used to create and prepare AI models. Pre-handling is critical to ensuring high-quality information. Four stages of information pre-handling are used to make the interaction easier: **(1) data acquisition, (2) storage (3) pre-processing, and (4) final analysis features**. Our model's ability to learn is dependent on the quality and useable data that can be extracted from our information; thus, pre-handling our information before putting it into a machine is essential to the learning model.

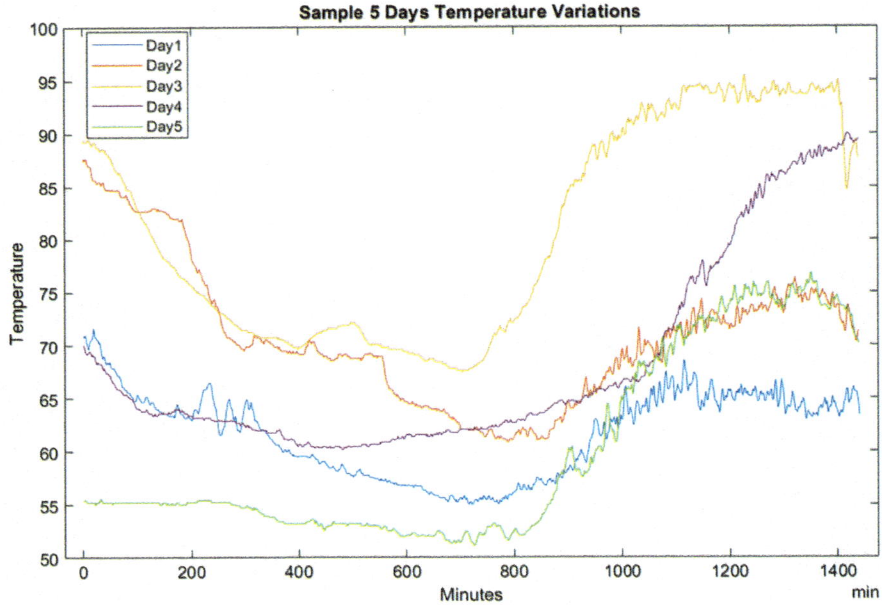

Fig. 4 Sample for temperature dataset

4.3 *Parameters of Performance*

Techniques have been evaluated using criteria such as accuracy, precision, recall, and F1 Score. This study's metrics are shown in Table 2. The confusion matrix displays the classified and misclassified clauses.

Table 2 Performance parameters

Name of metric	Equations		
Accuracy	$\text{Accuracy} = \frac{TP}{(TP+TN)*100}$		
Matrix of confusion		Positive	Negative
	Positive	True positive (TP)	False positive (FP)
	Negative	False negative (FN)	True negative (TN)
	$\textbf{Precision} = \frac{TP}{TP+FP}$		
	$\textbf{Recall} = \frac{TP}{TP+FN}$		
	$\textbf{F1 score} = \frac{2*\text{precision}*\text{recall}}{\text{precision}+\text{recall}}$		

5 Experiments and Discussion

The proposed approach is initially deployed on 3000 sample food datasets to demonstrate its effectiveness. The results are then compared to several cutting-edge algorithms. The information on the datasets and tuned parameters is offered first, followed by the test results, and lastly, the comments. All of the results are run on the same PC with the following specifications Software Specification: Windows 10 and Programming Language: MATLAB R2019a and Hardware Specification contains Ram 16G, CPU i7-4500, and hard disk 1 TB. All experiments were carried out using Matlab, and the results were extracted and analyzed.

The accuracy and ROC curve scores for several machine learning models are shown in Table 3 and Fig. 5. The Accuracy Score column indicates how many fresh and non-fresh observations were properly identified. The Accuracy Score column indicates how many fresh and non-fresh observations were properly identified. The accuracy and ROC Curve scores for several machine learning models are shown below.

Due to changes in data, the accuracy of each classifier fluctuates while using balancing techniques on a dataset. Fig. 6 clarifies the classification of the dataset, whether fresh or not, according to the dataset's size and measures each algorithm's

Table 3 The results of accuracy and ROC for different models

Machine learning model	Accuracy (%)	ROC score
Random forest optimizer	94.01	0.956
Random forest	89.32	0.951
Decision tree classifier	91.43	0.953
KNN classifier	90.31	0.959

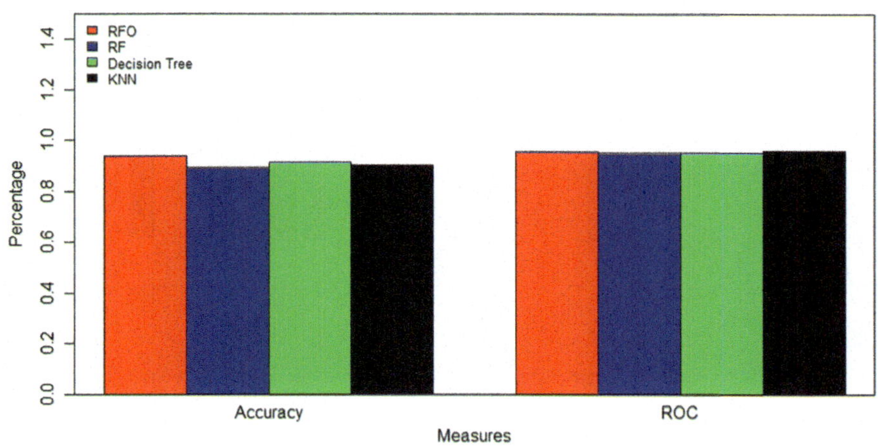

Fig. 5 Measures of accuracy and ROC

performance in the following Fig. 7, which shows accuracy, F1, precision, and recall metrics for each algorithm. The outcomes and comparisons of all approaches show that the random forest optimizer has demonstrated good accuracy of 94.01%.

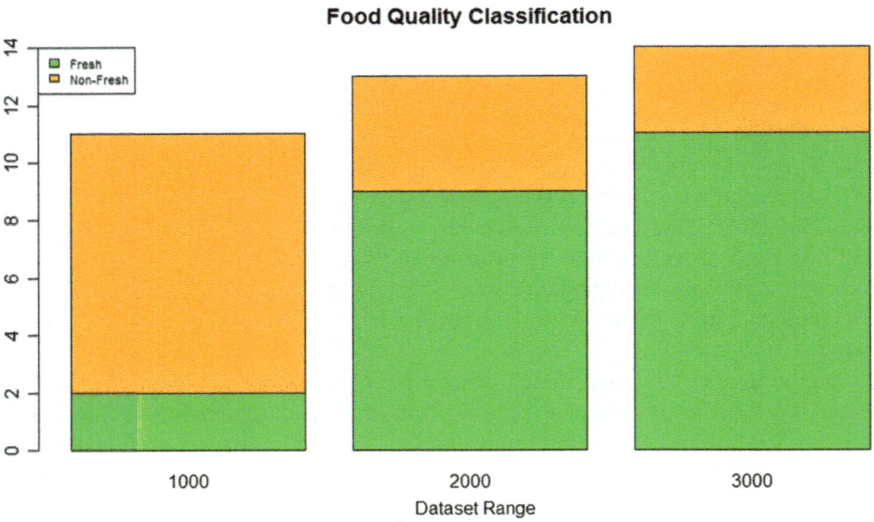

Fig. 6 Classification of proposed according to dataset size

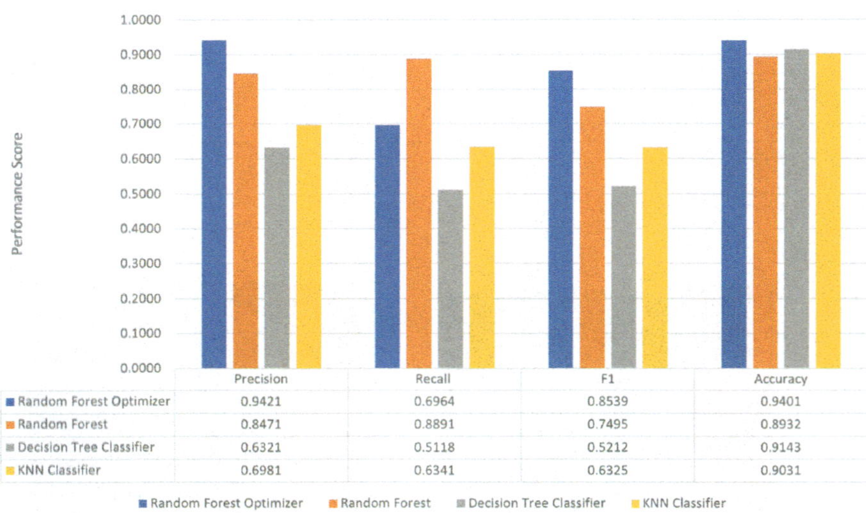

	Precision	Recall	F1	Accuracy
Random Forest Optimizer	0.9421	0.6964	0.8539	0.9401
Random Forest	0.8471	0.8891	0.7495	0.8932
Decision Tree Classifier	0.6321	0.5118	0.5212	0.9143
KNN Classifier	0.6981	0.6341	0.6325	0.9031

Fig. 7 Performance percentage of all classifiers in terms of accuracy, precision, F-measure, and recall metric

A food monitoring system is required to monitor the status of the food where it is stored to keep it from spoiling. The food monitoring system protects food from infection. The Random Forest Optimizer prediction-based food monitoring system is created, which monitors environmental parameters that affect food quality, such as temperature and humidity. This technology will assist in determining whether the environment in which food is stored is in good or terrible condition. The prediction of fresh or non-fresh food is based on the temperature and humidity levels in the storage environment. Existing classifiers and the proposed Random Forest Optimizer, i.e., Random Forest Optimizer, are incapable of making correct predictions. However, based on the suggested Random Forest Optimizer results, it predicts more accurately than other classifiers. The results demonstrate that RFO is approximately more accurate than the other classifiers.

6 Conclusion and Future Work

This research combined the Internet of Things (IoT) and Machine Learning technologies to precisely predict the state of food if food is fresh or non-fresh to reduce food waste. To do this, a low-cost variant of a standard e-nose was built and utilized to detect staleness in food products. The gas concentration was mapped and recorded for different ripening stages, and a data set was successfully recorded using a fruit. Machine learning is optimized to overcome existing limitations in the RF algorithm. Compared to the original RF Technique, it provides an optimization algorithm, the Random Forest Optimizer (RFO) Algorithm, that takes into account the uneven voting approach, where weight is assigned based on the performance of each tree. (RFO) has been used successfully in a variety of machine learning tasks. The random forest optimizer is used in this paper. Prediction for forecasting food quality, fresh or non-fresh by sending signals to the appropriate devices that are deployed to maintain suitable conditions. The suggested random forest optimizer performs around 94.01% accuracy better than other classifiers. Temperature and humidity are considered in this method to keep the atmosphere healthy. Aside from the fragrance, color and other factors must be addressed. This might be thought of as a potential future scope of this research.

References

1. K. Paliwal, S.R.N. Reddy, Smart food tracker smart monitoring system for food safety. In: National Conference on Product Design (NCPD 2016) (2016)
2. M. Maksimovic, V. Vujovic, E. Omanovic-Miklicanin, Application of Internet of Things in food packaging and transportation (2015). https://www.researchgate.net/publication/296486930
3. A. Venkatesh, T. Saravanakumar, S. Vairamsrinivasan, A. Vigneshwar, M. Santhosh Kumar, A food monitoring system based on bluetooth low energy and Internet of Things. Int. J. Eng. Res. Appl., **7**(3), 30–34 (2017). www.ijera.com

4. D. Sivakumar, K. Janaki, Effective food preservation from atmospheric conditions between harvest to retail marketing using IoT. Int. J. Technol. Res. Innov. Solutions (IJTRIS) **1**, 1–6 (2017)
5. S. Pler, M. Wolff, W.-J. Fischer, Food intake monitoring: an acoustical approach to automated food intake activity detection and classification of consumed food. Fraunhofer Institute for Photonic Microsystems (IPMS), Dresden, Germany, Institute of Acoustics and Speech Communication (IAS), Dresden University of Technology, Dresden, Germany (2012)
6. D.-Y. Lim, Y.-J. Ryoo, Development of remote monitoring system for cold-storage. In: 30th Annual Conference of IEEE Industrial Electronics Society, IECON 2004 (2004) 9, ed. by. W. Meulebroeck, H. Thienpont, H. Ottevaere, Photonics enhanced sensors for food monitoring: part 1. IEEE Instrum. Measur. Magaz. **19**(6), 35–45 (2016)
7. R. Jedermann, M. Nicometo, I. Uysal, W. Lang, *Reducing Food Losses by Intelligent Food Logistics* (The Royal Society Publishing, London, 2014)
8. M. Dowlati, S.S. Mohtasebi, M. Omid, S.H. Razavi, M. Jamzad, M. la Guardia de, Freshness assessment of gilthead sea bream (Sparus aurata) by machine vision based on gill and eye color changes. J. Food Eng. Elsevier Ltd. **119**, 277–287 (2013)
9. S. Sanahuja, M. Fédou, H. Briesen, Classification of puffed snacks freshness based on crispiness-related mechanical and acoustical properties. J. Food Eng., Elsevier
10. F. Angerosa, L.D. Giacinto, R. Vito, S. Cumitini, Sensory evaluation of virgin olive oils by artificial neural network processing of dynamic head-space gas chromatographic data. J. Sci. Food Agric. **72**, 323–328 (1996)
11. T. Rumpf, A.K. Mahlein, U. Steiner, E.C. Oerke, H.W. Dehne, L. Plümer, Early detection and classification of plant diseases with Support Vector Machines based on hyperspectral reflectance. Comput. Electron. Agricult., Elsevier BV **74**, 91–99 (2010)
12. N. El-Bendary, E. El Hariri, A.E. Hassanien, A. Badr, Using machine learning techniques for evaluating tomato ripeness. Expert Syst. Appl. **42**, 1892–1905 (2015). https://doi.org/10.1016/j.eswa.2014.09.057[CrossRef][GoogleScholar]
13. I.A. Basheer, M. Hajmeer, Artificial neural networks: fundamentals, computing, design, and application. J Microbiol Methods. **43**, 3–31 (2000). https://doi.org/10.1016/s0167-7012(00)00201-3[PubMed]
14. P. Cortez, A. Cerdeira, F. Almeida, T. Matos, J. Reis, Modeling wine preferences by data mining from physicochemical properties. Decis. Support Syst. Elsevier BV **47**, 547–553 (2009)
15. Y. Yang, C. Fu, Y.W. Chen, D.L. Xu, S.L. Yang, A belief rule based expert system for predicting consumer preference in new product development. Knowl.-Based Syst. **94**, 105 (2016)
16. E. Abad et al., RFID smart tag for traceability and cold chain monitoring of foods: demonstration in an intercontinental fresh fish logistic chain. J. Food Eng. **93**(4), 394–399 (2009). https://doi.org/10.1016/j.jfoodeng.2009.02.004
17. Sivakumar, D., Janaki, K.: Effective food preservation from atmospheric conditions between harvest to retail marketing using IoT. Int. J. Technol. Res. Innov. Solutions (IJTRIS) **1**, 1–6 (2017)
18. A.A.A. Al-Marakeby, F.A. Salem, Fast quality inspection of food products using computer vision. Int. J. Adv. Res. Comput. Commun. Eng. **2**, 4168–4171 (2013)
19. S.R. Dubey, A.S. Jalal, Apple disease classification using color, texture and shape features from images. Signal Image Video Process. **10**, 819–826 (2016)
20. Z. Yu et al. Hybrid adaptive classifier ensemble. IEEE Trans. Cybern. **45**, 177–190 (2015)
21. J.J. Huang et al., Fast image interpolation via random forests. IEEE Trans. Image Process. **24**, 3232–3245 (2015)
22. M. Maksimovic, V. Vujovic, E. Omanovic-Miklicanin, Application of Internet of Things in food packaging and transportation (2015)
23. S. Päßler, M. Wolff, and W. J. Fischer, Food intake monitoring: an acoustical approach to automated food intake activity detection and classification of consumed food. Physiol. Meas. **33**(6): 1073 (2012)
24. S.P. Heenan, N. Hamid, J.-P. Dufour, W. Harvey, C.M. Delahunty, Consumer freshness perceptions of breads, biscuits and cakes. Food Qual. Prefer. Elsevier Ltd **20**, 380–390 (2009)

25. S. Péneau, E. Hoehn, H.R. Roth, F. Escher, J. Nuessli, Importance and consumer perception of freshness of apples. Food Qual. Prefer. **17**, 9–19 (2006)
26. R.K. Megalingam, G.S. Sree, G.M. Reddy, I.R.S. Krishna, L. Suriy, Food spoilage detection using convolutional neural networks and k means clustering. In 2019 3rd International Conference on Recent Developments in Control, Automation and Power Engineering (RDCAPE) (Noida, IEEE, 2019), pp. 488–493
27. G.C. Green, A. D. Chan, R.A. Goubran, Monitoring of food spoilage with electronic nose: potential applications for smart homes. In 20093rd International Conference on Pervasive Computing Technologies for Healthcare (London, IEEE, 2009), pp. 1–7
28. L. Breiman, Bagging predictors. Mach. Learn. **24**(2), 123–140 (1996)
29. L. Breiman, Random forests. Mach. Learn. **45**(1), 5–32 (2001)
30. M. M. Ahmed, A. Taha, A. E. Hassanien, E. Hassanien. An optimized k-nearest neighbor algorithm for extending wireless sensor network lifetime. In International conference on advanced machine learning technologies and applications,Springer, Cham pp. 506–515 (2018, February)
31. M.M. Ahmed, E. H. Houssein, A. E. Hassanien, A. Taha, E. Hassanien. Maximizing lifetime of large-scale wireless sensor networks using multi-objective whale optimization algorithm. Telecommun. Syst. **72**(2), 243–259 (2019)
32. C. Yang, X.-C. Yin, Diversity-based random forests with sample weight learning. Cogn. Comput. **11**(5), 685–696 (2019)
33. P. Latinne, O. Debeir, C. Decaestecker, Limiting the number of trees in random forests, in: Lect. Notes Comput. Sci. (Including Subser. Lect. Notes Artif. Intell. Lect. Notes Bioinformatics), (Springer, 2001), pp. 178–187. https://doi.org/10.1007/3-540-48219-9_18.
34. A. Mellor et al., The performance of random forests in an operational setting for large area sclerophyll forest classification. Remote Sens. **5**(6), 2838–2856 (2013)
35. M. Dorigo, M. Birattari, in Ant Colony Optimization, ed by. C. Sammut, G.I. Webb., Encyclopedia of Machine Learning. (Springer, Boston, MA, Springer US, 2010), pp. 36–39
36. H. Liu, Y. Lei, Toward integrating feature selection algorithms for classification and clustering. IEEE Trans. Knowl. Data Eng. **17**(4), 491–502 (2005)

The manufacturer's authorised representative in the EU is Springer Nature Customer Service Centre GmbH, Europaplatz 3, 69115 Heidelberg, Germany. If you have any concerns regarding our products, please contact ProductSafety@springernature.com

Printed and bound by CPI Group (UK) Ltd, Croydon, CR0 4YY
28/09/2026
02224077-0015